Designing Control Strategies for Harmonic Power Filters to Improve the Power Quality in Distribution Networks

Harmonic Power Filters Design

RIVER PUBLISHERS SERIES IN POWER

Series Editors:

MASSIMO MITOLO
Irvine Valley College, USA

VIJAY K. SOOD
Ontario Tech University, Canada

The "River Publishers Series in Power" is a series of comprehensive academic and professional books focussing on the theory and applications behind power generation and distribution. The series features content on energy engineering, systems and development of electrical power, looking specifically at current technology and applications.

The series serves to be a reference for academics, researchers, managers, engineers, and other professionals in related matters with power generation and distribution.

Topics covered in the series include, but are not limited to:

- Power generation;
- Energy services;
- Electrical power systems;
- Photovoltaics;
- Power distribution systems;
- Energy distribution engineering;
- Smart grid;
- Transmission line development.

For a list of other books in this series, visit www.riverpublishers.com

Designing Control Strategies for Harmonic Power Filters to Improve the Power Quality in Distribution Networks

Harmonic Power Filters Design

Kumar Reddy Cheepati

KSRM College of Engineering, India

T. Mariprasath

KSRM College of Engineering, India

Marco Esteban Rivera Abarca

University of Nottingham, UK and Universidad de Talca, Chile

M. Nageswara Rao

UCEK (A), JNTUK, Kakinada, India

River Publishers

Routledge
Taylor & Francis Group
NEW YORK AND LONDON

Published 2025 by River Publishers
River Publishers
Broagervej 10, 9260 Gistrup, Denmark
www.riverpublishers.com

Distributed exclusively by Routledge
605 Third Avenue, New York, NY 10017, USA
4 Park Square, Milton Park, Abingdon, Oxon OX14 4RN

Designing Control Strategies for Harmonic Power Filters to Improve the Power Quality in Distribution Networks / by Kumar Reddy Cheepati, T. Mariprasath, Marco Esteban Rivera Abarca, M. Nageswara Rao.

Routledge is an imprint of the Taylor & Francis Group, an informa business

ISBN 978-87-7004-732-6 (hardback)
ISBN 978-87-4380-078-1 (paperback)
ISBN 978-87-7004-734-0 (online)
ISBN 978-87-7004-733-3 (master ebook)

Contents

Preface

As the world becomes increasingly reliant on electronic devices and non-linear loads, power quality has emerged as a critical concern for utilities, industries, and individuals alike. The proliferation of equipment such as variable frequency drives, switch-mode power supplies, and renewable energy systems has led to a significant increase in harmonic distortion, voltage sag, and other power quality issues. These disturbances can result in equipment malfunction, reduced lifespan, and even complete system failure.

In recent years, the use of shunt active and passive harmonic filters has emerged as an effective solution for mitigating these power quality problems. By compensating for harmonic currents and voltage disturbances, these filters can improve the reliability, efficiency, and overall performance of electrical systems.

This book provides a comprehensive guide to the design, implementation, and application of shunt active and passive harmonic filters for power quality improvement. With a focus on both theoretical and practical aspects, this book is an essential resource for engineers, researchers, and students working in the field of power quality and harmonic filtering. Through the use of case studies, examples, and simulation results, this book demonstrates the effectiveness of these filters in improving power quality and enhancing the reliability of electrical systems.

List of Figures

List of Tables

List of Abbreviations

AFHCC	Adaptive-fuzzy hysteresis current controller
AHCC	Adaptive hysteresis current controller
APF	Active power filter
APLC	Active power-line conditioner
BW	Bandwidth
CSI	Current source inverter
DTPF	Double-tuned passive filter
EQP	Electric power quality
FCS	Finite control set
FFT	Fast fourier transform
FOU	Foot point of uncertainty
FPGA	Field programmable gate array
HCC	Hysteresis current controller
HVDC	High voltage DC converter
IHD	Individual harmonic distortion
LOM	Largest of maximum
LPF	Low-pass filter
MOM	Middle of maximum
MPC	Model predictive controller
NB	Negative big
NM	Negative medium
NS	Negative small
PB	Pass band
PB	Positive big
PB	Positive big
PCC	Point of common coupling
PM	Positive medium
PM	Positive medium
PS	Positive small
PS	Positive small
PVB	Positive very big
PVS	Positive very small
QTPF	Quadruple-tuned passive filter

RAF	Ripple attenuation factor
RTDS	Real time digital simulator
SAF	Shunt active filter
SHAF	Shunt hybrid active filter
SMPS	Switched mode power supply
SOM	Smallest of maximum
SRF	Synchronous reference frame
STPF	Single tuned passive filter
SVC	Static VAR compensator
SVM	Space-vector-modulation
TCCC	Triangular carrier current controller
TPF	Tuned passive filter
TTPF	Triple-tuned passive filter
UPQC	Unified power quality conditioner
UPS	Uninterruptable power supply
VSI	Voltage source inverter
ZE	Zero

1

Introduction to the Power Quality and Harmonic Power Filters

1.1 Introduction

The production, transmission, distribution, and consumption of electrical power are the three primary areas that power system researchers often concentrate on. The electric power quality (EPQ) is specified by the power allocation, which is a dynamic connection between the production and consumption of power at the appropriate voltage and frequency. EPQ is widely used to express power delivery quality, supply consistency, and voltage and current quality. (i) Electrical loads that have non-linear characteristics and (ii) electrical loads interfering with telecommunication lines are the two main causes of power quality pollution. Lightning, notches, sags/swells, imbalanced voltage and current, interruptions, and harmonic distortions are the primary causes of electric power quality degradation. The distribution system's most important component is the EPQ. The main reason why the distribution system's power quality is poor is harmonics generation. Waveforms that are sinusoidal and have frequencies that are integral multiples of the power frequency are called harmonics. Both voltage and current distortion are referred to as harmonics. The fifth harmonic frequency is 250 Hz, the seventh harmonic frequency is 350 Hz, when the fundamental frequency is 50 Hz or 60 Hz, and so on. Harmonic generation is mostly caused by the exponential application of non-linear loads. All electronic power loads produce harmonics and are non-linear. Grid voltage distortion results from these loads' combination with the grid's impedance and non-sinusoidal current draw. Modern power electronic approaches offer ways to mitigate issues with power quality.

One of the most important concerns for both utilities and customers is power quality. The power system's performance is impacted by a number of power quality problems, including voltage sags, swells, interruptions, notches, flickers, transients, and harmonics. The majority of power quality problems are caused by non-linear loads, such as induction motors, bridge rectifiers, lift elevators, changeable speed drives, fluorescent lights, arc furnaces, etc. Harmonics are the most common power quality problem in power systems, and they can cause circuit breakers to malfunction, damage delicate equipment, overheat neutral conductors, copper and iron loss, thermal strains in cables, transformers, and rotating machinery, among other problems.

Filtering can be done passively or actively to improve power quality. Although passive filtering is a very effective and economical way to remove lower-order current harmonics, there are a few known disadvantages, including reliance on source impedance, losses, size, resonance, and compensating flexibility. For a comparatively little increase in size and expense, active filtering can remove all of these disadvantages. Source voltages and load currents have been sensed in order to extract the reference currents in SAF using time and frequency domain approaches.

Active filters operate best when they inject anti-harmonics into the distributed network, which allows them to filter various localized harmonic currents in a directed manner. Using traditional passive filters is not an option for this approach. Moreover, resonances brought on by the impedance of the surrounding electrical network and poor tuning from the dispersion of their characteristic parameters are two issues that can be avoided with the employment of active filters. A key component of the power system's safe operation is harmonic mitigation within the bounds of IEEE/IEC standards.

Any device that uses harmonic compensation must adhere to IEEE-519 voltage and current harmonic distortion restrictions. In this book, SAF and TPF have been designed using the most recent IEE-519-2022 standard.

1.2 Motivation

In general, various natures of linear and non-linear loads are connected to grid. The loads which are non linear in nature are the main causes of generating harmonics in distribution network. The induced harmonics in distribution network leads to various unwanted problems like transformers heating, poor power factor, torque thump in motors, resonance, drop in voltage due to harmonics through impedance, lowly operation of distribution network and even disturbs remaining sensitive loads coupled to the same point of common coupling (PCC).

Conventionally, TPFs were being used to alleviate the harmonics in the distribution network. TPFs are series connections of inductive and capacitive elements and are designed to tune particular lower-order current harmonics.

The whole TPF is in parallel with loads and is connected to PCC. The drawbacks of TPFs include fast ageing, problems in tuning, resonance, size, and fixed harmonic mitigation. Various topologies of static VAR compensators (SVCs) were being proposed to overcome the drawbacks of TPFs. Unfortunately, the SVC has an excessively slow response time and generates self-lower-order harmonics. Active power filters (APFs) or active power-line conditioners (APLCs) have been employed recently to compensate reactive power and mitigate harmonics. APF is connected either in series or in shunt and also both (series and shunt) as well as hybrid (combination of APF and TPF). The cost and kVA size of APF alone is more and it is not viable for low and medium voltage consumers. Connecting TPFs to APFs will reduce the cost and kVA size of APF and also improves the overall performance. The Shunt hybrid active filter (SHAF) is a combination of APF and TPF. SHAF connects in shunt with the non linear loads. It is the most commonly used configuration to mitigate harmonics in the distribution networks.

1.2.1 Harmonic sources

- Electronic and magnetic fluorescent lamps, arc furnace, power electronic drives, lift elevators, and switched mode power supplies (SMPSs).
- Power electric converters, controlled and uncontrolled rectifiers, inverters, SVCs, and cycloconverters.
- Single-phase power electronic loads such as uninterruptable power supplies (UPSs), computers, fax machines, Xerox machines, TVs, ovens, air conditioners, and refrigerators.
- Harmonic amplification resulting from series and parallel resonance.
- Malfunctioning of entire plant or part of the plant.
- Undesired functioning or failure of sensitive loads.
- Excessive heating and even failure of motors and generators.
- Excess loading, hotness and even failure of capacitors.
- Excess loading and hotness of transformers and neutral conductors.
- Errors in measurement in metering equipment.
- Frequent blow out of fuses, frequent tripping of circuit breakers.
- Glitches in voltage in computer leads to loss of data.
- TV, FM radio, communication, telephone systems, and signal processing units produce electromagnetic interference.

1.2.2 End effects of harmonics

- Burden on power

- Catastrophe of office equipment and exploitation of data or loss of data
- Workstation and organization lockups
- Increased down time of plants and increased cost of products
- No satisfaction of customer and less revenue

1.2.3 Mitigation of harmonics

The mitigation of harmonics is inevitable due to their severe end effects. The harmonic filters classification is as shown in Figure 1.1. The APF is connected either in series or in shunt and also both (series and shunt) as well as hybrid (combination of APF and TPF). The voltage is controlled by the series APF, which also separates the harmonics between the source and loads. In order to mitigate voltage sags and swells at the load side, the series APF can be regarded as a regulated variable voltage source because it produces voltage in phase with the supply voltage. Additionally, in order to maintain pure sinusoidal voltage across the load, the series APF generates harmonic

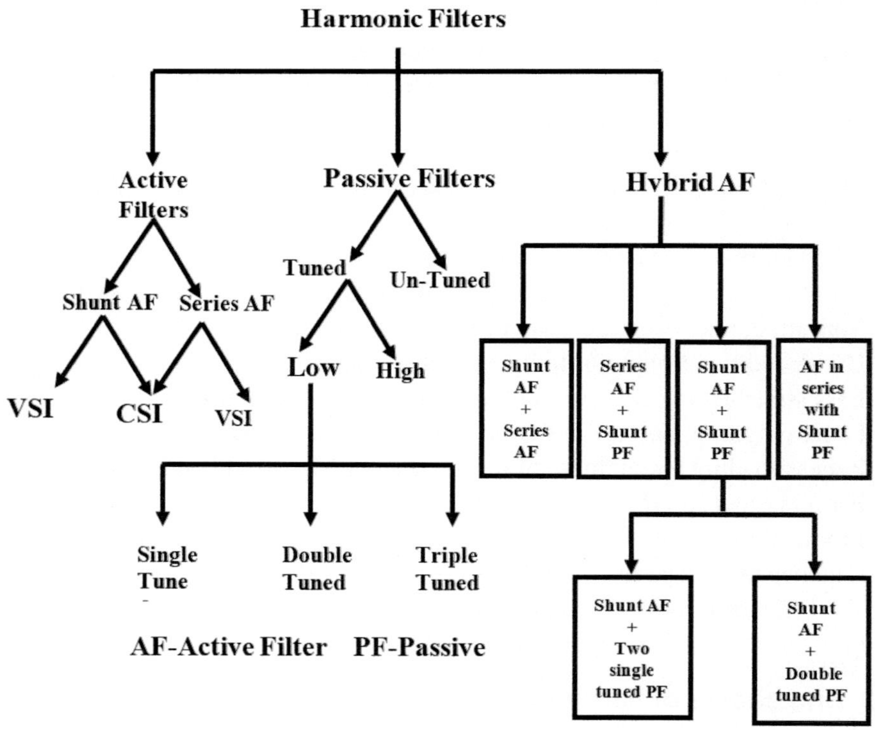

Figure 1.1 Harmonic filter classification.

voltages in quadrature with the source voltage. Hybrid APF is a mixture of TPFs and APFs. The hybrid series APF acts as a harmonic isolator between the sources and loads. The unified power quality conditioner (UPQC) is the combination of both series and shunt APF.

At the utility PCC, the series APF is capable of performing harmonic compensation and voltage regulation. Shunt active filters (SAFs) mitigate current harmonics, balance reactive power, rectify power factor, and regulate the voltage of the VSI capacitor. The extensive use of non-linear loads causes increased current harmonic distortion on the distribution side. The literature indicates that the SHAF is a suitable option for addressing the issues of excess reactive power and current distortion. The SHAF is also capable of power factor correction.

1.3 Shunt Active Filter

In essence, the SAF is a voltage source inverter (VSI) that produces harmonic currents that are 180 degrees out of phase with the harmonics of the load current. Control station and VSI make up SAF. The purpose of the control station is to inject harmonic currents that oppose the harmonics of the load current.

Harmonics in voltage and current are detected by the control station, which then counterclockwise injects them into the source. Its feedback-based closed-loop control technique reduces current distortion and produces a clean waveform for the loads. In addition to reactive power balancing and harmonic suppression, SAF may also perform voltage regulation and power factor correction.

The advantages of using SAF are

- Harmonic mitigation
- Elimination of neutral current
- Less strain on the distribution transformer
- Unity power factor
- Zero voltage regulation
- Automatic operation with changes in system condition
- Eradicating the danger of resonance between SAF and system impedance

1.4 Tuned Passive Filters

TPFs are widely used to mitigate lower-order current harmonics. Lower-order current harmonics are having high magnitude and hence dangerous in nature.

In general, TPFs operate based on resonance principle; at resonance TPFs offer lower impedance path for the flow of lower-order current harmonics which are tuned to that harmonic frequency and offer high impedance for the power frequency [18]. TPFs are classified as follows:

- ► Single-tuned passive filters (STPFs)
- ► Double-tuned passive filters (DTPFs)
- ► Triple-tuned passive filters (TTPFs)
- ► Quadruple-tuned passive filters (QTPFs)

Usually, STPFs and DTPFs are most widely in distribution network to eliminate lower-order current harmonics. At resonance, DTPF offers lower impedance path for two tuned lower-order current harmonic frequencies with a single circuit and circuit breaker per phase, whereas STPFs require two circuits with two circuit breakers per phase, hence the cost and size of DTPF is less as compared with two STPFs [18]. The performance of DTPF in mitigating the lower-order current harmonic is more as compared with two STPFs. STPFs and DTPFs are connected in parallel with the SAF. This topology is called as shunt hybrid active filter (SHAF) as shown in Figure 1.2. The advantage of SHAF is that, TPFs takes the responsibility of mitigating the lower-order current harmonics whereas SAF is responsible for mitigating higher-order current harmonics. Thus the overall cost and kVA size of SAF is reduced when TPFs are connected in parallel to it at the PCC. In this book, if the SAF is connected to two single-tuned passive filters per

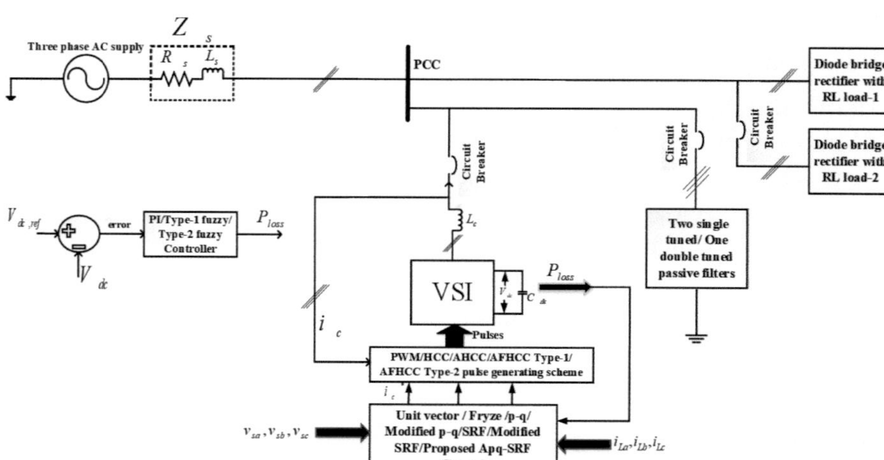

Figure 1.2 Block diagram of the proposed research.

phase then it is termed as single-tuned SHAF and if the SAF is connected to one double-tuned passive filter per phase then it is termed as double-tuned SHAF.

1.5 Block Diagram and Contributions of the Book

The proposed book's detailed block diagram is displayed in Figure 1.2. With two non-linear loads, the SAF and TPF are coupled in a shunt configuration. Two uncontrolled diode bridge rectifiers with R-L load are used as non-linear load, one load is on at $t = 0$ and another load is on at $t = 0.5$ sec. Thus, it forms like a dynamic non-linear load. Initially, the SHAF is tested with the existing reference current extraction techniques such as unit vector, p-q, modified p-q, SRF, and modified SRF techniques in conjunction with two single-tuned passive filters/one double-tuned passive filter, various pulse generating schemes such as TCCC/HCC/AHCC/AFHCC-1/AFHCC-2 and PI/Type-1 fuzzy/Type-2 fuzzy logic DC-link voltage controllers. The drawbacks of all existing reference current extraction techniques, DC-link voltage controllers, tuned passive filters, and pulse generating schemes are discussed in Chapters 2 and 3. The comparative analysis is presented in Chapter 4, and hardware results are presented in Chapter 5. A novel reference current extraction technique that is, Averaging pq-SRF (Apq-SRF) technique is proposed and implemented to compensate the drawbacks of the existing techniques. Under both constant and variable loading scenarios, the suggested Apq-SRF technique is contrasted with a number of current reference current extraction methods, tunable passive filters, pulse generating systems, and DC-link voltage controllers. The MATLAB Simulink results revealed that the proposed Apq-SRF technique is working well in conjunction with tuned passive filters and pulse generating schemes under constant and dynamic loading conditions as compared with available reference current extraction schemes. Further, the proposed Apq-SRF technique is implemented in SAF hardware with TMS5700-XL processor in conjunction with PI DC-link voltage controller and HCC pulse generating scheme. The suggested Apq-SRF technique can reduce current harmonics in accordance with IEE-519-2022 standards, according to the hardware results. Chapter 5 compares the simulation findings with the hardware results. This work can be considered a future scope because the performance of the suggested Apq-SRF approach is not tested under dynamic loading conditions due to practical restrictions and component availability.

1.6 Approaches used in the Book

In this book, single-tuned SHAF and double-tuned SHAF is chosen as a configuration to investigate the various reference current extraction techniques, pulse generators, DC-link voltage regulators and TPFs. The SHAF is connected in shunt with two diode bridge rectifiers at PCC for harmonic compensation. PWM-built two-level VSI operates as a current control source and two STPFs or one DTPF is used in association with SAF. To extract the reference currents, fast-responding time domain control techniques were employed. The proposed SHAF generate anti current harmonics so that the source current turn into sinusoidal, the SAF is used for cancel out the higher-order current harmonics whereas the TPFs are used for mitigating the lower-order current harmonics.

Figure 1.2 shows the detailed topology of SHAF with various SAF control techniques and TPFs. The DC side of VSI is connected to the storage capacitor and the inductor is connected to the ac-side. The grid is connected to all types of loads together with non-linear loads. The SAF operates as a feedback loop, the inverter is controlled to inject the current through the coupling inductor to convert source current sinusoidal. Thus, SAF can compensate reactive power, correct the power factor, and mitigate current harmonics.

The interfacing inductor offers isolation to transform the voltage signal to a current signal for mitigating current harmonics. The current control is limited by the switching frequency of VSI and also the voltage developed in the coupling inductor. The voltage in interfacing inductor governs the maximum *di/dt* that can be attained by SAF. The high *di/dt* values are essential to mitigate the harmonic currents of higher order. A high value of inductor is required for quarantine from the power system and also for protecting from oscillatory disturbances. But, the large value of inductor may bound the capability of the SAF to mitigate current harmonics of higher order.

For the harmonic current to be generated 180 degrees out of phase with the load current at PCC, SAF is necessary. The SAF's role should ideally be limited to compensating for the network's reactive power consumption.

As a result, the average DC capacitor voltage should remain steady. But in reality, it wouldn't be consistent. The voltage of the DC capacitor decreases as a result of the VSI and coupling inductor losing power.

However, the DC-link capacitor voltage should remain near constant so that the SAF system will not get affected. The constant DC-link voltage is obtained by supplying active power from the grid to compensate the losses by the closed loop mechanism. The following points need to be discussed to know the importance of the SHAF.

- The efficient method for reference current extraction
- The feedback control mechanism for both transient and steady state
- The power converter for the power circuit
- The indirect PWM-pulse generators
- Efficient design of STPFs and DTPF

1.7 Literature Reviews

The SAF controller and design of TPFs are the most important part of the SHAF and currently various SAF control strategies and TPFs are reported in the literature. The following SAF controllers and TPFs are required for SHAF configuration.

(i) Fast responding time domain reference current extraction techniques
(ii) Control methodologies for pulse generation
(iii) The VSI capacitor voltage regulators
(iv) Design of single- and double-tuned TPFs

1.7.1 Reference current extraction techniques

For reference current extraction, the literature suggests a number of control strategies. In the literature, time and frequency domain approaches were mostly being offered. The FFT analysis serves as the foundation for frequency domain methods. Frequency domain approaches have the primary benefit of being appropriate for both single- and three-phase systems. Among the frequency domain methods are the modified Fourier series method and the sine multiplication method [19, 20]. Techniques in the time domain are developed for instantaneous current signals that are obtained from harmonic current signals. Time domain methods are primarily applied to four-wire or three-phase three-wire systems. Response times for time domain approaches are faster than those for frequency domain techniques. Fryze suggested a voltage-and-current-based control method [21]. Using Clarke transformations, Akagi developed a SAF for a three-wire, three-phase system based on instantaneous reactive power theory [22]. Extension for three-phase systems with zero-sequence components is also feasible. Using the instantaneous load voltages and currents, the reference currents are retrieved instantly and without delay [22]. The reference currents-based park transformation technique was created by Bhattacharya. Synchronous reference frame theory is the term for this method [23]. Watanabe created a hypothesis that applies to general voltage and current signals in both steady-state and dynamic states [24]. Nassar Mendalek and Peng A. presented the instantaneous reactive power theory, which may be used to three-phase power systems that are balanced or unbalanced, sinusoidal or non-sinusoidal, and have zero-sequence currents or not [25]. According to Reyes S's comparative analysis of vectoral theory, p-q theory, d-q theory, modified or cross-product concept, and p-q-r reference frame, all of these approaches achieved their goals when the grid was optimal. A nonlinear control method for obtaining the SAF reference currents was created by Salem Rahmani [26]. Compensation for reactive, unbalanced, and harmonic load current components is offered by this method. Rondineli presented a novel method for enhancing the harmonic detection for SAF's transient reaction time [27]. Ricardo put out a reliable control method for non-linear load balancing, harmonic compensation, and power-factor adjustment [28].

1.7.2 PWM-current control technique

There were several SAF switching control strategies in use, with a focus on PWM-VSI approaches. David Broad Kazmierkowski provides a general overview of PWM-current controllers in a three-phase load [29]. The employment of predictive controllers, the most intricate and costly technology, may restrict the distribution network's dynamic reaction. Although ramp comparison controllers require a big gain and a lower bandwidth, they have the advantage of being able to lower the maximum PWM-inverter switching frequency and produce distinct harmonics. Although the hysteresis current controller with three independent controls functions well, it has the disadvantage of not functioning effectively when there are few cycles and when the VSI switching frequency is higher than intended. Switching frequency can be decreased by introducing zero volts at the right periods. The proposal by Akira Nabae is a unique feedback control. Both in the steady state and in the transient state, it can solve the quick current response problem and suppress higher order harmonic currents [30]. For the machine drive system, Bimal K. Bose proposes an adaptive hysteresis-band current control method of PWM-VSI [31]. The modulation frequency is maintained as a constant. Though only internal permanent magnet synchronous machine load is taken into consideration, this technique can be used to AC drives and loads of different kinds. Marian presented two new methods for controlling the PWM-VSI current controller. Based on three-level hysteresis comparators, both of these techniques switch electrically programmable read-only memory tables to choose the proper VSI output voltage vectors. The first controller operates using current components defined within a stationary reference system. The second one makes use of a system of rotating DC components. Three distinct two-level hysteresis controllers are discussed and compared with the fundamental idea behind these methods [29]. Dixon [32] evaluates various current modulation methods for VSI. He compared various carrier signals and the performance was evaluated. The best harmonic distortion mitigation and reduced current ripple have been demonstrated by the triangular carrier controller. This method's overshoot issues via proportionate gain are its drawback. The periodic controller performs effectively when slow power switches and time delays are used. The ease of use of the hysteresis current control technique makes it extremely popular. Introducing a new hysteresis current controller based on space-vector-modulation (SVM) [33] is Bong-Hwan Kwon. It makes use of both the SVM controller and the hysteresis current controller. It controls a set of space vectors identified by a region detector and applies the selected space vector to the main hysteresis current controller. It employs a set of space vectors, including the zero vector, to minimize switching frequency. These vectors are determined based on the output frequency sign and the signals from three comparators, utilizing a wider hysteresis band compared to the standard hysteresis controller.

G.H. Bode has published a three-level hysteresis current controller for a single-phase inverter [34]. In comparison to the two-level hysteresis current controller, it achieves a decrease in magnitude and switching frequency fluctuation and enhances efficiency. Murat Kale introduces an adaptive hysteresis band current controller for SAF [35]. It adjusts the hysteresis bandwidth according to the modulation frequency, supply voltage, DC-link capacitor voltage, and reference compensator current slope. However, because of the high frequency, it has more switching losses, which can be addressed by the adaptive-fuzzy-hysteresis current controller that Hamoudi proposed [36].

B. Mazari presents parametric optimization of SAF with fuzzy hysteresis control [37]. It is employed to generate the switching signals and determine the ideal decoupling inductance value. By employing the instantaneous grid voltage and current as input variables and the hysteresis band as an output variable, this method enables the definition of a systematic hysteresis band and the creation of a fuzzy rule table. It keeps the modulation frequency constant.

For optimal SAF performance, all control strategies are now built in DSP/FPGA processors and coded in high-level languages. The PWM approach used with VSI must maintain a consistent DC voltage and satisfy the load's harmonic requirements.

1.7.3 DC-link voltage controllers

DC-link voltage controllers have been used to maintain the constant DC voltage across the VSI capacitor bank and to maintain the energy balance between the source and load. Traditional controllers like PI, PID, Fuzzy, PI-Fuzzy, and other possible solutions such as dead-beat control, SM control (SMC), adaptive control, resonant control, and repetitive-based control have been used for this purpose. In this research, the following controllers are used.

- ✓ A PI controller
- ✓ Fuzzy controller of type 1 [38].
- ✓ Fuzzy controller of type 2 [38].

1.7.4 Tuned passive filters

Tuned passive filters (TPFs) are used to mitigate the lower-order current harmonics and it gives relief to SAF from eliminating lower-order current harmonics. The cost and kVA size of SAF will reduce by adding TPFs in parallel to it [39]. In this research, the following TPFs are used.

- ✓ Single-tuned passive filter
- ✓ Double-tuned passive filter using equivalent and parametric method

1.8 Objectives of the Book

From the literature reviews, it is clear that there is a need to develop a robust reference current extraction technique which is ideal for all supply and load conditions in distribution networks.

The main goal of this proposed study is to develop and apply a unique reference current extraction method for SAF and design TPFs in accordance with the IEE-519-2022 standard to reduce harmonic distortion.

Exponential growth of non-linear loads leads to increase in current harmonic distortion in the distribution system. The SHAF offers distribution networks unity power factor, reactive power correction, and current harmonic mitigation. Various techniques were being suggested in the literature. Moreover, existing techniques exhibit several disadvantages. By observing the above-mentioned deliberations, the objectives of the book are listed as follows:

- ✠ To design VSI-based SAF that delivers sufficient bandwidth, achieve good efficiency, and transient performance.
- ✠ To develop a novel reference current extraction technique for SAF that eliminates the drawbacks of existing techniques.
- ✠ To implement efficient pulse generation technique for generating switching pulses to the IGBT/MOSFET VSI-based SAF according to the control reference.
- ✠ Through hardware and simulations, the SAF's performance in terms of power factor correction, reactive power compensation, and harmonic mitigation is compared with different controllers and system settings.
- ✠ To design various single- and double-tuned passive filters for effective operation of the SHAF.
- ✠ To develop SAF hardware setup and validate with the simulation results.

1.9 Scopes and Contributions

This book's main contribution is the design and implementation of the SAF employing both the current reference current approaches and a new Apq-SRF reference current extraction method. The suggested Apq-SRF technique's performance is evaluated against that of the current reference methods, pulse generation schemes, and DC-link voltage regulators. The following is a breakdown of the job scopes.

- The SAF is built using PWM-VSI (two-level). The suitable PWM-VSI technique is projected to power circuit for triggering SAF switches.
- For practical implementation, TMS57004-XL DSP controller is used which is a simple and cost-effective solution.
- Two uncontrolled diode rectifiers are used as non-linear loads.

To separate the reference currents from the polluted load current, a number of management strategies were being developed. For SAF, the reference current extraction methods listed below have been studied.

- ► Unit vector control
- ► Fryze power theory
- ► Instantaneous real and reactive power (p-q) theory
- ► Modified p-q theory
- ► Synchronous reference frame (SRF) theory

- ► Modified SRF theory
- ► Proposed Averaging pq-SRF (Apq-SRF) control

The following indirect PWM-VSI techniques were being designed to diminish the switching power losses.

- ► Current controller with a triangular carrier
- ► Current controller with constant hysteresis
- ► Hysteresis-adaptive current controller
- ► Type 1 and Type 2 adaptive-fuzzy-hysteresis current controllers

For the VSI capacitor bank to have a consistent DC voltage, DC-link voltage regulators are necessary. To keep the DC voltage across the VSI capacitor bank steady, the following controllers were created. These fuzzy logic controllers are Type-1, Type-2, and PI controllers.

1.10 Organization of the Book

This book is well organized into six chapters. The brief description of each chapter is as follows:

The idea of SAF and TPFs is briefly explained in Chapter 1. A thorough literature analysis and the significance of harmonic mitigation are explained. The book's goals and chapters are also described.

Chapter 2 enlightens various single- and double-tuned passive filters, configuration, design techniques, advantages, and drawbacks.

The SAF topologies, harmonic properties, fundamentals of compensation, and design are covered in Chapter 3. VSI is used to establish the SAF topologies. The SAF is categorized into single-phase and three-phase systems to accommodate various non-linear loads. Additionally, it explains the different reference current extraction methods, including the proposed Apq-SRF control theory, unit vector control theory, fryze power theory, p-q theory, modified p-q theory, SRF theory, and modified SRF theory. Additionally, it describes the various indirect PWM-VSI control methods needed for SAF. The two-level PWM-inverter's necessary switching pulses are produced by the TCCC, fixed-HCC, adaptive-HCC, and adaptive-fuzzy-HCC approaches. The type-2 fuzzy HCC approach reduces switching losses while improving performance.

Chapter 4 examines how the single- and double-tuned SHAFs operate by examining different SAF control strategies and TPF design. Each of the reference current extraction methods with several PWM control methods is verified independently through extensive MATLAB simulation. Under both steady and transient diode bridge rectifier loads, the SHAF is confirmed. The suggested methods and TPFs were found to be superior by the THD measurement.

Chapter 5 explains the general conclusions and future scope which is an outcome from this book.

1.11 Summary

In this chapter, concise preamble of the power quality issues is presented. The importance of harmonic mitigation, harmonic effects, and benefits of harmonic mitigation are highlighted. The objectives, scopes, and contributions of the book are listed. Organization of the book has been given at the end of this chapter.

2

Tuned Passive Harmonic Power Filter Design

2.1 Introduction

This chapter presents the design of various tuned passive harmonic filters along with their merits and demerits in each case. In brief, passive filters are classified into tuned filters and untuned filters. Mostly tuned passive filters are used along with the SAF to reduce the cost and kVA size of the SAF. In this proposed research, tuned passive filters are connected in parallel with the SAF at the PCC, thus it forms a shunt hybrid active filter (SHAF) configuration. The main advantage of tuned passive filters is their ability to effectively eliminate lower-order current harmonic frequencies by being tuned to their resonance frequency, while the remaining harmonics are addressed by the SAFs [40]. Broadly tuned passive filters are classified into:

- Single-tuned passive filters
- Double-tuned passive filters
- Triple-tuned passive filters
- C-type filter
- Second-order high-pass filters

2.2 Design of Single-tuned Passive Harmonic Filter

A series RLC circuit tuned to a single harmonic frequency that typically offers a low harmonic impedance characteristic is called a single-tuned passive filter. Its overall impedance may be found using

$$Z_n = R_n + j\left(\omega L_n - 1/\omega C_n\right). \tag{2.1}$$

The frequency at resonance condition $\omega L_n = 1/\omega C_n$.
From eqn (2.1) $Z_n = R_n$.

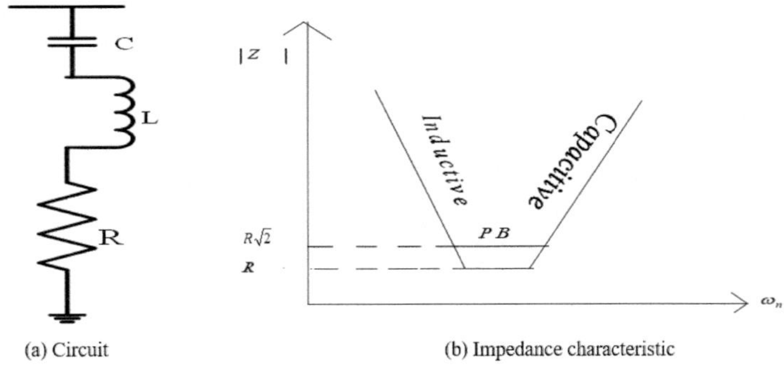

Figure 2.1 Single-tuned filter (a) circuit (b) impedance characteristic.

At a tuned frequency, the capacitive and inductive reactances should be equivalent for a perfect filter. A filter's sharpness of tuning is measured by its quality factor (Q), which determines whether it is a high or low type. The value ranges from 30 to 60, and the high-type filter is tightly tuned to one of the lower harmonic frequencies. A filter with a low Q-factor, typically in the range of 0.5 to 5.5, exhibits low impedance across a wide frequency range and is commonly known as a high-pass filter. For a tuned filter, the Q-factor is defined as the ratio of inductance (or capacitance) to resistance at the resonant frequency. The quality factor (Q) is given by

$$Q = X/R. \tag{2.2}$$

The frequencies at which the filter reactance equals its resistance, or when the impedance angle is 45°, and the magnitude is as indicated in Figure 2.1, are the boundaries of the filter pass band (PB).

The quality factor and pass band relationship might be stated as follows:

$$Q = \omega_n/PB. \tag{2.3}$$

2.2.1 Tuned passive filters

The following relationships can be used to calculate the filter impedance in terms of δ and Q:

$$\omega = \omega_n\left(1 + d\right). \tag{2.4}$$

$$\omega = \frac{1}{\sqrt{L.C}}. \tag{2.5}$$

At the tuned frequency the inductance and capacitive reactance are equal and is given by

$$X_0 = \omega_n L = \frac{1}{\omega_n C} = \sqrt{\frac{L}{C}}. \tag{2.6}$$

$$Q = \frac{X_0}{R}. \tag{2.7}$$

$$C = \frac{1}{\omega_n X_0} = \frac{1}{\omega_n RQ}. \tag{2.8}$$

$$L = \frac{X_0}{\omega_n} = \frac{RQ}{\omega_n}. \tag{2.9}$$

Substituting eqn (2.4), (2.8), and (2.9) into eqn (2.1) yields

$$\omega L = \omega_n \left(1 + \delta\right) . \frac{RQ}{\omega_n} = \left(1 + \delta\right) RQ. \tag{2.10}$$

$$\frac{1}{\omega C} = \frac{1}{\omega_n \left(1 + \delta\right) . \frac{1}{\omega_n RQ}} = \frac{RQ}{\left(1 + \delta\right)}. \tag{2.11}$$

$$Z_f = R\left\{1 + jQ\delta\left(\frac{2 + \delta}{1 + \delta}\right)\right\}. \tag{2.12}$$

Taking into account that δ is very small in comparison to unity

$$Z_f R\left(1 + j2\delta Q\right) = X_0\left(Q^{-1} + j2\delta\right). \tag{2.13}$$

$$Z_f R(1 + 4\delta^2 Q^2)^{1/2}. \tag{2.14}$$

The tuning factor (δ) indicates the degree of filter detuning from the nominal tuned frequency. The following factors contribute to this filter detuning: (i) changes in the fundamental frequency; (ii) aging or temperature-induced changes in the filter capacitance and inductance; and (iii) initial off-tuning brought on by manufacturing tolerances and finite tuning step sizes. The entire detuning is displayed in units of the nominal tuned frequency by eqn (2.15).

$$\delta = \left(\omega - \omega_n\right) / \omega_n. \tag{2.15}$$

Furthermore, a 2% shift in L or C results in the same detuning as a 1% shift in system frequency. Consequently, d is frequently written as

$$\delta = \frac{\Delta f}{f} + \frac{1}{2}\left(\frac{\Delta L}{L} + \frac{\Delta C}{C}\right). \tag{2.16}$$

Table 2.1 Anticipated errors.

Items	Anticipated errors
Frequency	Δf:2%
Differences in inductance (L)	ΔL:−10≈+20
A change in capacitance (C)	ΔC:−4.5≈+20

Table 2.2 Tuning in an applied tuning factor.

Order	5th	7th	9th	11th	13th	15th	17th	19th
Tuning order	4.813	6.734	8.663	10.59	12.51	14.44	16.36	18.29

Equation (2.17) illustrates the connection between the ideal and actual tuning orders.

$$h = h_n + (\delta h_n)\,, \tag{2.17}$$

where h is the tuning order considered tuning factor (δ) and h_n is the optimum tuning order. According to Table 2.1 and eqn (2.16), the optimal tuning factor (δ) is 0.0375.

Each harmonic's tuning order is displayed in Table 2.2 and is determined using eqns (2.16), (2.17), and Table 2.1 displays the expected errors.

2.2.2 Current absorption rate of single-tuned filter

The filter side receives the majority of the harmonic current if the filters are precisely tuned at the harmonic frequency. That is, approximately I_{5f}/I_a = 95.7% is the harmonic current absorption in a fifth-order filter. The rate of harmonic current absorption would, however, drop significantly to 70.17% if the filter capacitance were raised to +6.5% over the intended range. When the remaining 30% of the harmonic current leaves the power source, it negatively impacts nearby systems or equipment. Additionally, the harmonic current absorption rate would drop to 83.87% if the filter's capacitance were reduced to −4.5% of its intended value. In this instance, the power supply also experiences a 20% harmonic current flow.

2.2.3 Quality factor

There is a scale that indicates the degree of harmonic current absorption amount and the quality factor (Q), which establishes the sharpness of tuning [41, 42]. In this regard, depending on their intended usage, all filters will be

Table 2.3 Filter absorption effects.

Filter		Current (I_f)	Absorption rate (%)
5th filter	Original	4.785	95.7
	X_C + 5%	3.5.9	70.17
	X_C − 4.5	4.194	83.87
	X_L + 20	1.475	29.49
	X_L − 10%	2.598	51.95
7th filter	Original	4.873	97.46
	X_C + 5%	3.938	78.76
	X_C − 4.5	4.350	87.00
	X_L + 20	1.984	39.68
	X_L − 10%	3.244	64.88

Table 2.4 Current harmonic content after installation.

Filter	$Q_5(\%)$	$Q_{10}(\%)$	$Q_{30}(\%)$	$Q_{50}(\%)$	$Q_{70}(\%)$	$Q_{100}(\%)$
5th	28.6	51.2	87.28	94.48	97.98	98.66
7th	38.54	64.12	92.90	97.98	98.58	99.38

either high or low Q types.

$$Q = \frac{f_r}{f_{HI} - f_{LO}} = \frac{f_r}{3dB}, \text{ or } Q\,(series) = \omega_0 \frac{2\omega_n}{P_{loss}} = \frac{\omega_n L}{R} = \frac{1}{\omega_n RC}. \tag{2.18}$$

Higher Q, lower R. Q is dependent on R. Q and bandwidth have an inverse relationship. As a result, Q determines the frequency selectivity of filters. High Q, for instance, can precisely pick frequency and has a tiny BW.

The quality factors are 5, 10, 30, 50, 70, and 100, respectively, to assess the filter efficiency. In Table 2.4, the current absorption rates are displayed.

2.3 Design of Double-tuned Passive Filter

The standard double-tuned filter is depicted in Figure 2.2. It is made up of two resonant circuits, one in series and one in parallel, coupled in series. At series resonance frequency (w_s), the series resonant circuit has low impedance, while the parallel resonant circuit has high impedance at parallel resonance frequency (w_p). As seen in Figure 2.3, the two resonant frequencies can have a low impedance channel thanks to the combination of a series and parallel resonance circuit. There are two ways to construct the double-tuned filter: parametric and equivalent methods.

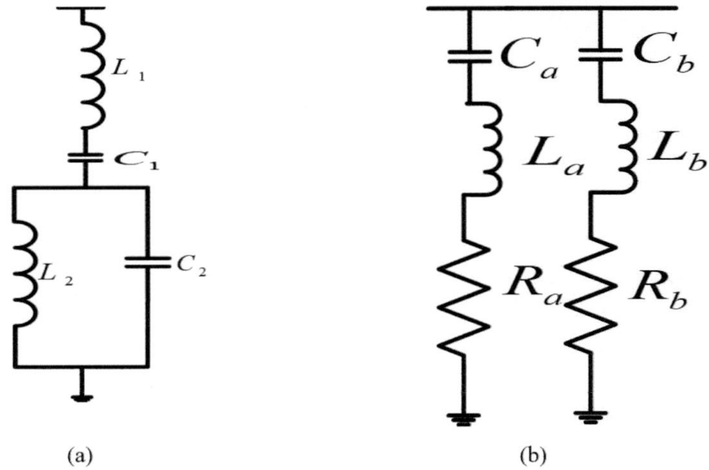

Figure 2.2 Double-tuned filter (a) circuit (b) equivalent single-tuned parallel filter.

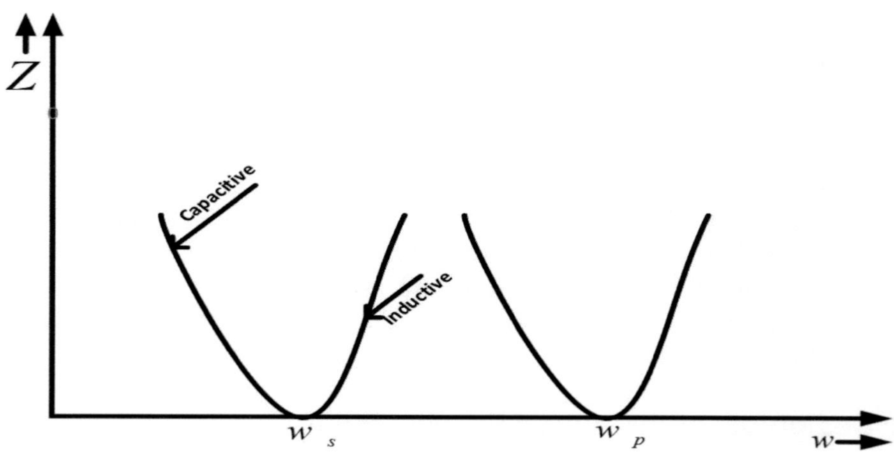

Figure 2.3 Characteristics of double-tuned passive filter.

2.3.1 Equivalent method of designing double-tuned filter

With this approach, two single-tuned filters connected in parallel can be used to determine the double-tuned filter parameters. Both the double-tuned filter and the two parallel single-tuned filters have identical impedances [43]. The corresponding double-tuned passive filter approach is taken into consideration in this proposed study, along with the suggested SAF configuration.

The series circuit impedance is

$$Z_s = j\omega L_1 + \frac{1}{j\omega C_1} \tag{2.19}$$

The parallel circuit impedance is

$$Z_p = \left(j\omega C_2 + \frac{1}{j\omega L_2}\right)^{-1}. \tag{2.20}$$

The total impedance is

$$Z = j\omega L_1 + \frac{1}{j\omega C_1} + \left(j\omega C_2 + \frac{1}{j\omega L_2}\right)^{-1}. \tag{2.21}$$

$$Z = \frac{\left(1 - \frac{\omega^2}{\omega_s{}^2}\right)\left(1 - \frac{\omega^2}{\omega_p{}^2}\right) - \omega^2 L_2 C_2}{Jv C_2 \left(1 - \frac{\omega^2}{\omega_p{}^2}\right)}. \tag{2.22}$$

The expression for the parallel resonance frequency (ω_s) and series resonance frequency (ω_p) in radians/sec is

$$\omega_s = \frac{1}{\sqrt{L_1 C_1}} \omega_p = \frac{1}{\sqrt{L_2 C_2}}. \tag{2.23}$$

Let, w_a, w_b are the resonance frequencies of two passive filters that are single-tuned.

$$\omega_a = \frac{1}{\sqrt{L_a C_a}} \omega_b = \frac{1}{\sqrt{L_b C_b}}. \tag{2.24}$$

The expression for the impedance of two parallel single-tuned filters is

$$Z_{ab} = \frac{\left(1 - \frac{\omega^2}{\omega_a{}^2}\right)\left(1 - \frac{\omega^2}{\omega_b{}^2}\right)}{j\omega c_a \left(1 - \frac{\omega^2}{\omega_b{}^2}\right) + j\omega c_b \left(1 - \frac{\omega^2}{\omega_a{}^2}\right)}. \tag{2.25}$$

Two passive filters that are single-tuned have the same cumulative impedance as a double-tuned filter.

$$Z = Z_{ab}. \tag{2.26}$$

Comparing coefficient of ω^4

$$\omega_a \omega_b = \omega_s \omega_p. \tag{2.27}$$

Comparing coefficient of ω

$$C_1 = C_a + C_b. \tag{2.28}$$

Comparing coefficient of ω^3

$$C_b \frac{1}{\omega_a^2} + C_a \frac{1}{\omega_b^2} = C_1 \frac{1}{\omega_p^2}. \tag{2.29}$$

The parameter L_1 is given by

$$L_1 = \frac{1}{C_a {\omega_a}^2 + C_b {\omega_b}^2}. \tag{2.30}$$

The double-tuned filter's parallel resonance frequency ω_s and series resonance frequency ω_p can be found by

$$\omega_s = \frac{1}{\sqrt{L_1 C_1}}, \tag{2.31}$$

$$\omega_p = \frac{\omega_a \omega_b}{\omega_s}. \tag{2.32}$$

Since, ω_a is the zero of double-tuned filter impedance, so Z (ω_a) = 0. The equation to solve L_2 is

$$\left(1 - \frac{{\omega_a}^2}{{\omega_s}^2}\right)\left(1 - \frac{{\omega_a}^2}{{\omega_p}^2}\right) - \omega^2 L_2 C_1 = 0. \tag{2.33}$$

The eqn (2.33) can be simplified to get L_2

$$L_2 = \frac{\left(1 - \frac{{\omega_a}^2}{{\omega_s}^2}\right)\left(1 - \frac{{\omega_a}^2}{{\omega_p}^2}\right)}{C_1 {\omega_a}^2}. \tag{2.34}$$

The value of C_2 can be obtained by

$$C_2 = \frac{1}{L_2 {\omega_p}^2}. \tag{2.35}$$

Therefore, the parameters of two parallel-connected single-tuned filters (L_a, C_a, L_b, C_b) may be used to determine all the parameters required for a double-tuned filter $(L_1, C_1, L_2,$ and $C_2)$.

Compared to two single-tuned passive filters, this double-tuned passive power filter can perform better at removing lower-order current harmonics. Compared to single-tuned passive filters, double-tuned passive filters are smaller and need less reactive power. Therefore, the optimum option for removing two lower-order current harmonics is a doubletuned passive filter [43].

2.3.2 Parametric method of designing double-tuned filter

The disadvantage of the equivalent method is that it is not precise and does not account for the real reactive power demand of the non-linear loads. By employing existing information, such as tuned frequencies, parallel resonant frequencies, and reactive power demands of non-linear loads, the parameters of a double-tuned filter can be computed using the parametric approach [44].

A double-tuned filter has zero overall impedance at tuned frequencies.

$$Z = j\omega L_1 + \frac{1}{j\omega C_1} + \left(j\omega C_2 + \frac{1}{j\omega L_2}\right)^{-1} = 0. \tag{2.36}$$

Eqn (2.36) gives the following relation

$$\omega^4 L_1 L_2 C_1 C_2 - \omega^2 \left(L_2 C_1 + L_1 C_1 + L_2 C_2\right) + 1 = 0. \tag{2.37}$$

Using Vida's theory, the roots of eqn (2.37). Vida's hypothesis states that the correlation between eqn (2.37)'s roots and coefficients is as follows:

$$\omega_1^2 \omega_2^2 = (L_1 C_1 L_2 C_2)^{-1}, \tag{2.38}$$

i.e.,

$$\omega_1 \omega_2 = \frac{1}{\sqrt{L_1 C_1}} \frac{1}{\sqrt{L_2 C_2}} = \omega_s \omega_p. \tag{2.39}$$

The tuned frequencies in eqn (2.39) are denoted by $\omega_1 \,\&\, \omega_2$, whereas the series and parallel resonant frequencies are denoted by $\omega_s \,\&\, \omega_p$. With the double-tuned filter, the harmonic currents at tuned frequencies ($\omega_1 \,\&\, \omega_2$) will be avoided. After defining the tuned frequencies, the parameters $L_1, C_1, L_2,$ and C_2 will be determined. The filter impedance is 0 at the series resonant frequency (ω_s) and infinite at the parallel resonant frequency (ω_p). When designing, engineers should consider the network harmonic impedance and use a parallel resonant frequency ω_p appropriately to avoid drastically amplifying certain harmonic currents close to ω_p. If $\omega_1, \omega_2,$ and ω_p are identified, then eqn (2.21) unquestionably defines ω_s. Equation (2.23) can be substituted into eqn (2.27) to obtain

$$\frac{\omega^4}{{\omega_s}^2 {\omega_p}^2} - \left(\frac{C_1}{C_2}\frac{1}{{\omega_p}^2} + \frac{1}{{\omega_s}^2} + \frac{1}{{\omega_p}^2}\right)\omega^2 + 1 = 0. \tag{2.40}$$

Since, ω_1, ω_2 are the solutions of eqn (2.40), substituting ω by $\omega_1 \& \omega_2$, the relationship between $C_1 \& C_2$ can be obtained as

$$\frac{C_1}{C_2} = \frac{{\omega_1}^2 + {\omega_2}^2 - {\omega_p}^2}{{\omega_s}^2} - 1. \tag{2.41}$$

From eqn (2.38), L_1 can be calculated as

$$L_1 = \left(\frac{\omega_p}{\omega_1\omega_2}\right)^2 . \frac{1}{C_1}. \tag{2.42}$$

From eqn (2.23), L_2 can be calculated as

$$L_2 = \frac{1}{{\omega_p}^2} * \frac{1}{C_2} = \frac{1}{{\omega_p}^2 C_1}\left(\frac{{\omega_1}^2 + {\omega_2}^2 - {\omega_p}^2}{{\omega_s}^2} - 1\right). \tag{2.43}$$

In addition to avoiding the harmonic currents, the double-tuned filter compensates for reactive power, which the non-linear load requires. Assuming that the fundamental system voltage is U and the filter's fundamental reactive power is Q, the filter's impedance at the fundamental frequency (ω_F) equals

$$Z\left(\omega_F\right) = \frac{-jU^2}{Q}. \tag{2.44}$$

The impedance of the filter at the fundamental frequency is

$$Z\left(\omega_F\right) = j\omega_F L_1 + \frac{1}{j\omega_F C_1} + \left(j\omega_F C_2 + \frac{1}{j\omega_F L_2}\right)^{-1}. \tag{2.45}$$

Equating eqn (2.44) and (2.45) gives the value of capacitor C_1

$$C_1 = \left\{\frac{1}{\omega_F} - \omega_F\left(\frac{\omega_p}{\omega_1\omega_2}\right)^2 + \frac{\omega_F\left[\left({\omega_1}^2 + {\omega_2}^2 - {\omega_p}^2\right){\omega_p}^2 - {\omega_1}^2{\omega_2}^2\right]}{{\omega_1}^2{\omega_2}^2\left({\omega_p}^2 - {\omega_F}^2\right)}\right\}\frac{U^2}{Q}. \tag{2.46}$$

Since ω_p is first established and $\omega_1, \omega_2,$ and Q are known, the parameters of the fundamental double-tuned filter may be calculated.

In case, when $\omega_s = \omega_p = \omega_0 = \sqrt{\omega_1\omega_2}$, the equations can be simplified as

$$\frac{C_1}{C_2} = \frac{\left(h_2 - h_1\right)^2}{\left(h_1 h_2\right)} \tag{2.47}$$

$$C_1 = \frac{{h_1}^2 + {h_2}^2 - \left(h_1 h_2\right)^2 - 1}{\left(h_1 h_2 - 1\right) h_1 h_2} . \frac{Q}{\omega_F U^2} \tag{2.48}$$

$$L_1 = \frac{1}{{\omega_0}^2 C_1} \tag{2.49}$$

$$L_2 = \frac{\left(\omega_1 - \omega_2\right)^2}{{\omega_0}^4 C_1}, \tag{2.50}$$

where, $h_1 = \frac{\omega_1}{\omega_F}$, $h_2 = \frac{\omega_2}{\omega_F}$.

2.3.3 Design of damped double-tuned filter

The drawback of the conventional double-tuned filter is that, in situations where the filter impedance is equal to the system impedance, the resonance condition greatly increases the harmonics [45]. A conventional double-tuned filter, also called a damped double-tuned filter, uses a resistor in an appropriate design to keep the harmonics from being unduly amplified during the resonance state, as shown in Figure 2.4.

As seen in Figure 2.4, there are numerous damped double-tuned filter designs available. The ultimate goal is to suppress the excessive voltage harmonics during the resonance state, regardless of the kind of damped double-tuned filter used.

In Figure 2.4 (a), when ω is greater than ω_2, the resistance must be an asymptotic line. The resistance R is set so that the inductive and capacitive reactances must cancel each other out to produce a high impedance channel during resonance conditions, so that Figure 2.4 (b) functions as a high-pass filter when ω is above ω_2. The filter's impedance at ω_p in Figure 2.4 (c) is of the order of R. The filter's impedance in Figure 2.4 (d) ought to correspond with the relationship of $R > \sqrt{L_1/C_1}$. Both at ω_p and above ω, the filter in Figure 2.4 (e) has limited impedance, and its parameters should match the relationship of $R > \sqrt{L_1/C_1}$. Finally, the filter of Figure 2.4 (f) with its

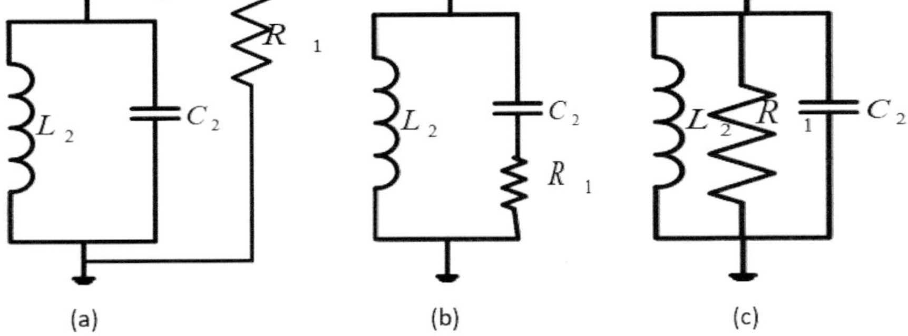

Figure 2.4 *Continued.*

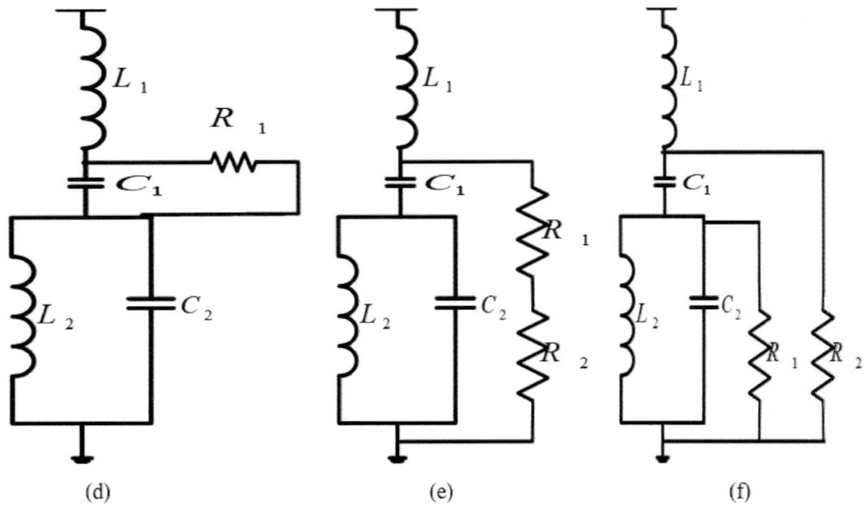

Figure 2.4 Various types of damped double-tuned passive filters.

impedances both at ω_p and above w_2 are limited to the lower value of R_1 and R_2.

2.3.4 Optimization of damped double-tuned filter

The damped double-tuned filter's impedance can be scanned in the frequency domain to reveal three pairs of tuned frequencies [45]. Figure 2.5 displays the algorithm for double-tuned filter optimization.

1. The frequencies ω_{Z1}, ω_{Z2}, at which the impedances of the filter are minimum;
2. The frequencies ω_{X1}, ω_{X2}, at which the total reactance of the filter equals zero.
3. Others are the ω_1 and ω_2 solutions where the filter's impedance is zero.

2.4 Two Inductances and Two Capacitors Paralleled Structure Double-tuned Filter

As illustrated in Figure 2.6, the two capacitances paralleled structured double-tuned filter and the two inductances paralleled structured double-tuned filter are shown in Figure 2.7. These filters are useful in lowering the current harmonics at the high voltage direct current (HVDC) converters.

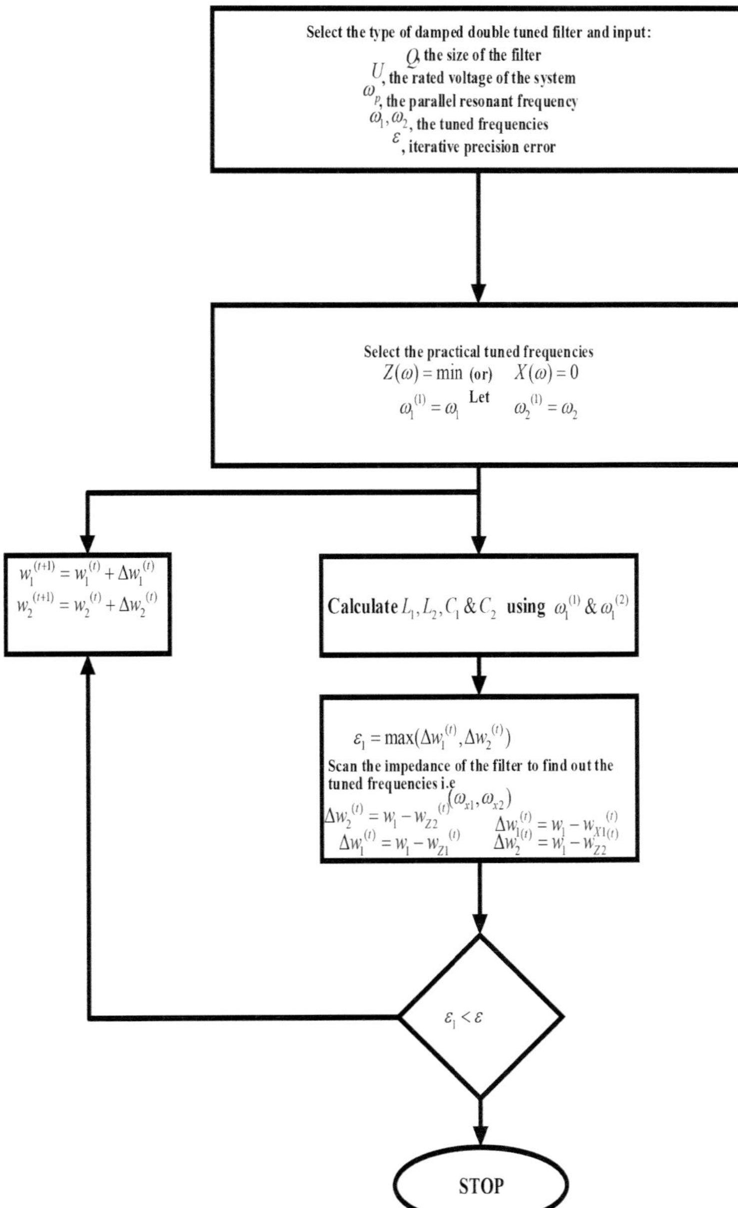

Figure 2.5 Flowchart for the optimization of damped double-tuned passive filter.

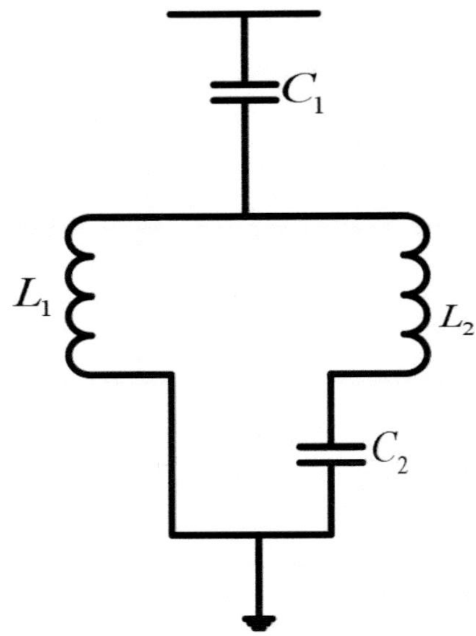

Figure 2.6 Double-tuned passive filter with two inductances paralleled structure.

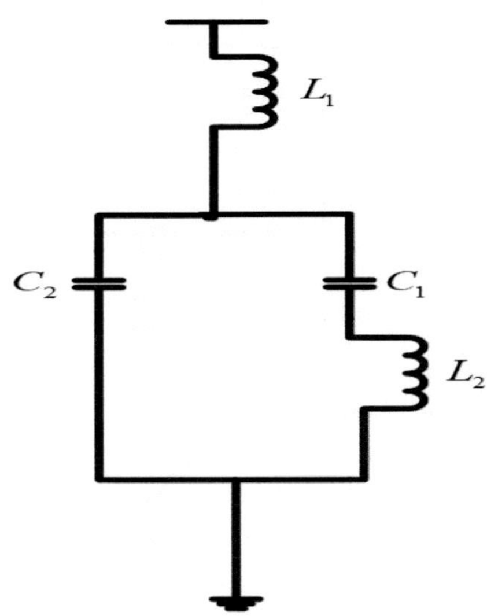

Figure 2.7 Double-tuned filter with two capacitances paralleled structure.

2.4.1 Parameter calculation of two capacitances and inductances paralleled double-tuned filters

Two approaches, the parametric method and the equivalent method, can be used to create the two inductance and capacitance paralleled structured double-tuned filter [46, 47].

2.4.1.1 Using equivalent method

The conventional double-tuned filter's impedance $Z(s)$ can be reduced to that of a parallel filter with two inductors using the following continuous fractions:

$$Z\left(S\right) = \frac{1}{SC_1} + \frac{1}{\frac{1}{S(L_1 + L_2)} + \frac{1}{\frac{(L_1 + L_2)^2}{SL_2^2 C_2} + \frac{1}{\frac{L_2}{SL_1(L_1 + L_2)}}}}. \tag{2.51}$$

In order for the parameters of a paralleled filter with two inductors to equal:

$$C_1' = C_1, \tag{2.52}$$

$$L_1' = \frac{L_1\left(L_1 + L_2\right)}{L_2}, \tag{2.53}$$

$$C_2' = \frac{L_2^2 C_2}{(L_1 + L_2)^2}, \tag{2.54}$$

$$L_2' = \left(L_1 + L_2\right). \tag{2.55}$$

The parameters of two capacitances parallel filters can be obtained similarly as follows:

$$L_1' = L_1 \tag{2.56}$$

$$C_1' = \frac{C_1^2}{C_1 + C_2} \tag{2.57}$$

$$L_2' = \frac{L_2(C_1 + C_2)^2}{C_1^2} \tag{2.58}$$

$$C_2' = \frac{C_1 C_2}{C_1 + C_2}. \tag{2.59}$$

2.4.1.2 Using parametric method

The total impedance of two inductors paralleled filter is as follows:

$$Z(S) = \frac{1}{SC_1} + \frac{SL_2\left(SL_1 + \frac{1}{SC_1}\right)}{SL_{2.}\left(SL_1 + \frac{1}{SC_2}\right)}. \tag{2.60}$$

When Z(*S*) = 0,

$$\omega^4 L_1 L_2 C_1 C_2 - \omega^2 \left(L_1 C_2 + L_2 C_2 + L_2 C_1\right) = 0. \tag{2.61}$$

When *Z(S)* is infinite, the circuit produced paralleled resonance suppose the frequency is ω_0, that is

$$\omega_0 = \sqrt{\frac{1}{C_2\left(L_1 + L_2\right)}}. \tag{2.62}$$

At the basic of frequency ω_F, assuming that the fundamental wave's rated voltage is U, the filter should also emit reactive power Q. The fundamental wave impedance of the filter is as follows [47]:

$$Z(\omega_F) = -J\frac{U^2}{Q} = -J\frac{\omega_F^2(L_1C_2 + L_2C_2 + L_2C_1) - \omega_F^4 L_1 L_2 C_1 C_2 - 1}{\omega_F^3(L_1 + L_2)C_1C_2 - \omega_F C_1}, \tag{2.63}$$

$$C_1 = \frac{Q\left(\omega_F^2\omega_0^2\left(\omega_1^2 + \omega_2^2\right) - \omega_F^4\omega_0^2 - \omega_1^2\omega_2^2\omega_0^2\right)}{U^2\left(\omega_F^2\omega_1^2\omega_2^2 - \omega_F\omega_0^2\omega_1^2\omega_2^2\right)}, \tag{2.64}$$

$$L_2 = \left(\frac{1}{\omega_1^2} + \frac{1}{\omega_2^2} - \frac{1}{\omega_0^2}\right)\frac{1}{C_1}, \tag{2.65}$$

$$L_1 = \frac{\omega_0^2 L_2}{L_2 C_1 \omega_1^2 \omega_2^2 - \omega_0^2}, \tag{2.66}$$

$$C_2 = \frac{1}{\omega_0^2\left(L_1 + L_2\right)}. \tag{2.67}$$

2.5 Results and Discussion

The comparison of tuned passive filters is as shown in Table 2.5. The double-tuned passive filter is efficient as compared with two single-tuned passive filters. The parametric method based double-tuned filter gives better

Table 2.5 Comparison of tuned passive filters.

	Individual harmonic distortion (IHD)(%)								
	3rd	**5th**	**7th**	**9th**	**11th**	**13th**	**15th**	**17th**	**19th**
Without any filter	0	22.04	10.05	0	7.20	4.71	0	3.17	2.33
With two single-tuned passive filters	0	5.04	3.02	0	5.21	3.92	0	2.16	1.45
With single double-tuned passive filter (equivalent method)	0	4.91	2.98	0	4.96	3.53	0	1.99	1.41
With single double-tuned passive filter (parametric method)	0	4.8	2.92	0	4.9	3.42	0	1.9	1.38
With single two capacitance paralleled double-tuned filter (equivalent method)	0	4.85	2.9	0	4.85	3.48	0	1.89	1.38
With single two capacitance paralleled double-tuned filter (parametric method)	0	4.75	2.85	0	4.8	3.38	0	1.85	1.3
With single two inductance paralleled double-tuned filter (equivalent method)	0	4.82	2.85	0	4.88	3.45	0	1.92	1.4
With single two inductance paralleled double-tuned filter (parametric method)	0	4.72	2.82	0	4.75	3.32	0	1.8	1.32

results as compared with equivalent method. Hence, in this book parametric method based double-tuned passive filter is used along with the shunt active filter. In this book, six pulse uncontrolled diode bridge rectifiers are used as non-linear load and the source is assumed to be balanced. Hence, no zero-sequence harmonics ($3n$ where 'n' is odd number) will generate from the system.

2.6 Contributions

- Single- and double-tuned passive filters are designed mathematically and implemented in MATLAB for under fixed non-linear loading conditions.
- Comparative study of single- and double-tuned passive filters has been done against their performance in mitigating lower-order current harmonics.
- The relative performance of parametric method-based double-tuned passive filter is more as compared with equivalent method-based double-tuned filter.

2.7 Summary

In this chapter, many tuned passive filters are presented, based on application, cost and size only suitable tuned passive filter can be used along with the SAF. In this proposed research, two single-tuned passive filters or parametric method-based double-tuned passive filter in each phase is used along with the SAF to eliminate lower-order current harmonics.

3

Shunt Active Harmonic Power Filters Design

3.1 Introduction

This chapter focuses on the SAF topologies, types of grid systems, harmonics classification and basic compensating, design of SAF system, various reference current extraction techniques, and pulse-generating schemes. Typically, the SAF is made to mitigate current harmonics in the AC grid. The SAF can be used for low-power ratings less than 100 kVA, medium-power ratings between 100 kVA and 10 MVA and high-power ratings of greater than 10 MVA. The SAF includes the control unit and the grid circuit. In general, grid circuits are developed with converters and can be categorized into

- Voltage source inverter (VSI)
- Current source inverter (CSI)

The SAF can be used for two types of grid systems to meet the demand of different loads which are non-linear in nature.

- One-phase grid system
- Three-phase grid system (three wire or four wire)

The SAF design consists of VSI, the capacitor which acts as a DC source, coupling inductor, and control system. By combining different reference current extraction techniques with pulse generating techniques, the control methodology is achieved. For real-world applications, SAFs are now implemented using DSP/FPGA/RTDS/D-Space.

The existing reference extraction approaches for the SAF are mostly divided into two categories: time domain and frequency domain techniques. The current reference is extracted from the line current harmonic using the time domain approach and basic algebraic computations. The most precise multiple and individual harmonic detection is offered by the frequency domain, which is based on the fast Fourier transform (FFT) procedure. The time domain method's applicability lies in its quicker response time when compared to the frequency domain. The time domain techniques listed below were created for the SAF. As mentioned in Chapter 1, a few methods can be discovered in the literature.

- Unit vector control theory
- Fryze power theory
- Instantaneous real and reactive power (p-q) theory
- Modified p-q theory
- Conventional synchronous reference frame (SRF) theory
- Modified SRF theory
- Proposed Average pq-SRF (Apq-SRF) control strategy

Based on PWM current control approaches, the SAF switches are pulsed using a variety of pulse-generating schemes. It is difficult to operate the current control effectively because of the unknown and fluctuating inverter load. The SAF must produce anti-harmonics in accordance with the load's requirements, as is widely known. It should be mentioned that the SAF system provides current in relation to the reference current. The characteristics of current control strategies for the SAF's effective operation are listed below.

- Enhanced use of VSI, for generating high current in a polluting load
- Less stationary and dynamic current-control errors
- Extensive linear variation range
- To reduce the output current harmonics, use short amplitude harmonics of lower order in the output voltage
- Small switching losses in VSI switches' power transistors
- Short workout duration and simple implementation

The direct PWM-VSI current control methods listed below are developed and applied for the SAF in this study.

- Triangular carrier current controller (TCCC)
- Controller for fixed-hysteresis current (HCC)
- Adaptive-HCC (AHCC)
- Adaptive-fuzzy-HCC – type-1 & type-2 (AFHCC type-1 & AFHCC type-2)

3.2 SAF Topologies

The SAF was primarily created for power factor correction in distribution systems, reactive power compensation, grid and load balancing, and harmonic mitigation. Either VSI or CSI configuration is supported by the SAF's design. To enhance the quality of the current, the CSI serves as a harmonic current source. VSI is the most convenient topology for the SAF and it behaves as a voltage source controlled by current.

3.2.1 Current source inverter

The plot from CSI is displayed in Figure 3.1 (a). It consists of six movable switches. It must supply all of the current that the load demands. The PCC is connected to the current source SAF by series transformers, which separate the inverter currents from the network's frequency components. The DC current grid uses a big DC inductor connected in series. The highest compensatory current should be at least equal to the inductor DC current. Direct output current control, excellent converter dependability because of the unaltered environment, and built-in protection are some of the source's special benefits. Nonetheless, it features a DC reactor that supports it and guarantees constant DC current distribution.

Once more, the CSI is more costly than VSI and requires a lot of capacitor-like filters at AC places to eliminate undesired current harmonics. The multilevel or cascaded architecture, which permits harmonic correction at greater power levels, cannot use this modification [48].

3.2.2 Voltage source inverter

Since the PWM-VSI is lightweight, inexpensive, and has been upgraded in multilevel topologies to enhance its performance of high-power balance correction at low rotation rate, it can be used in SAF applications [48]. The VSI's structure is displayed in Figure 3.1 (b). It is built using six-dimensional reduced power transistors with anti-parallel diodes. VSI acts as a regulated

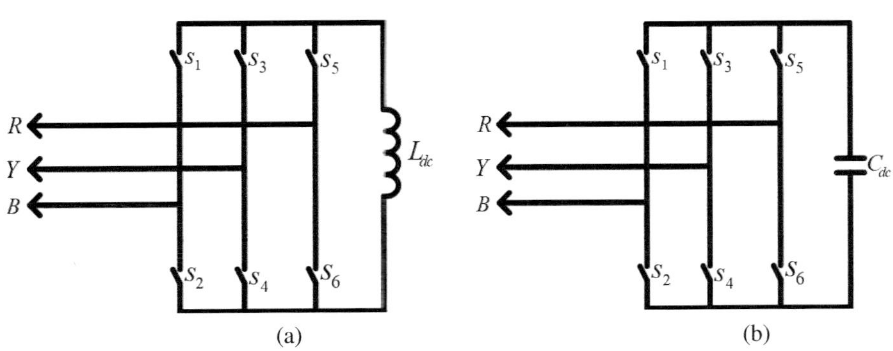

Figure 3.1 SAF topology (a) voltage source inverter (b) current source inverter.

source of current energy for electrical power. The assembly states that an inducer interface (transformer) was utilized to link the system to the ac amplifier using a dual-level VSI. The DC voltage is maintained by the capacitor. The goal of this kind of setup is to offset the non-linear load in the 100 kVA intermediate power range caused by the small number of semiconductors. Nevertheless, the PWM-VSI multilevel or multilayer developed SAF for high-power and intermediate applications.

3.3 SAF Configurations

For the purpose of accommodating various non-linear load types in grid systems, the SAF is essentially divided into a one-phase system and a three-phase system.

3.3.1 Single-phase system

SAF can be used to enhance the operation of distributed power systems due to harmonics in single-phase client equipment. A single-phase SAF can mitigate the associated issues caused by the extensive usage of computers, printers, electronics, offices, and household appliances [48]. A VSI with two power lines, a four-phase transformer, and a DC-link capacitor is used to implement a single phase SAF, as seen in Figure 3.2. The interface inductor connects the inverter to the PCC in accordance with the loads. The single-phase SAF eliminates the harmonics by delivering a present compensation that is equal

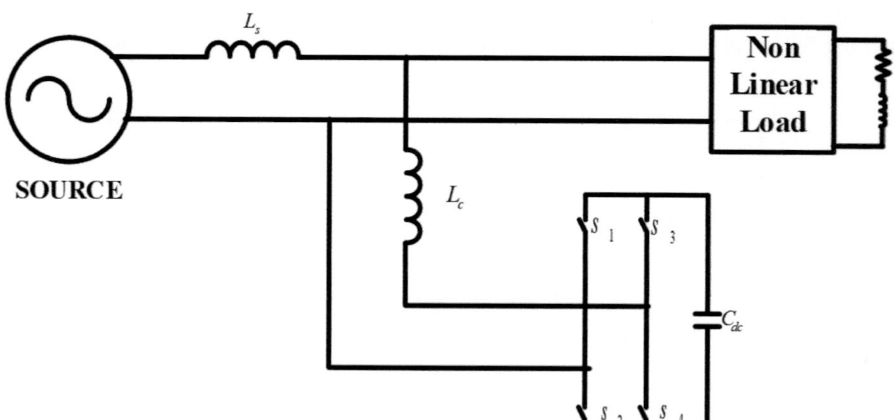

Figure 3.2 Single-phase shunt active filter system.

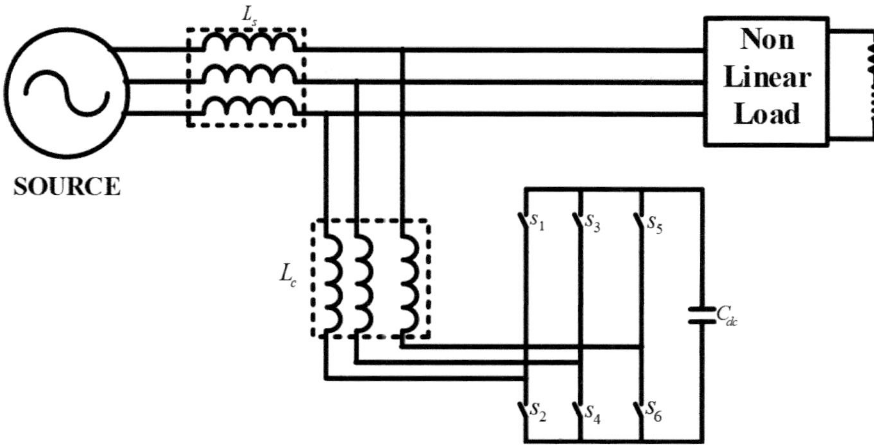

Figure 3.3 Three-phase three-wire shunt active power system.

but opposite [48]. The current is now sinusoidal and phased in voltage as a result.

3.3.2 Three-phase three-wire system

The two-level PWM-VSI has three-phase, three-wire SAFs, as seen in Figure 3.3. Six-power transistors, a DC-Link capacitor, and an interface inductor connect the PWM-VSI to the PCC. Because of the power transistor's switching function, the interface inductor suppresses high-order harmonic components [48]. By injecting identical non-harmonic components with the PCC, the original distortion is eliminated and the energy quality of the associated electrical system is improved due to the absence of current harmonics.

3.3.3 Three-phase four-wire system

Two configurations are available for the PWM-VSI with a three-phase, four-wire SAF: (i) a four-leg inverter, where the AC is connected to the fourth leg of the neutral inverter, and (ii) the three-leg inverter and the AC.

The DC-bus midway is directly connected to the neutral wire. The second strategy is straightforward in its design and reasonably priced. As seen in Figure 3.4, it makes use of a conventional three-leg inverter, in which the DC-link capacitor is divided and its middle is linked to the fourth wire, allowing for a return to the neutral current. This four-wire, three-phase SAF system

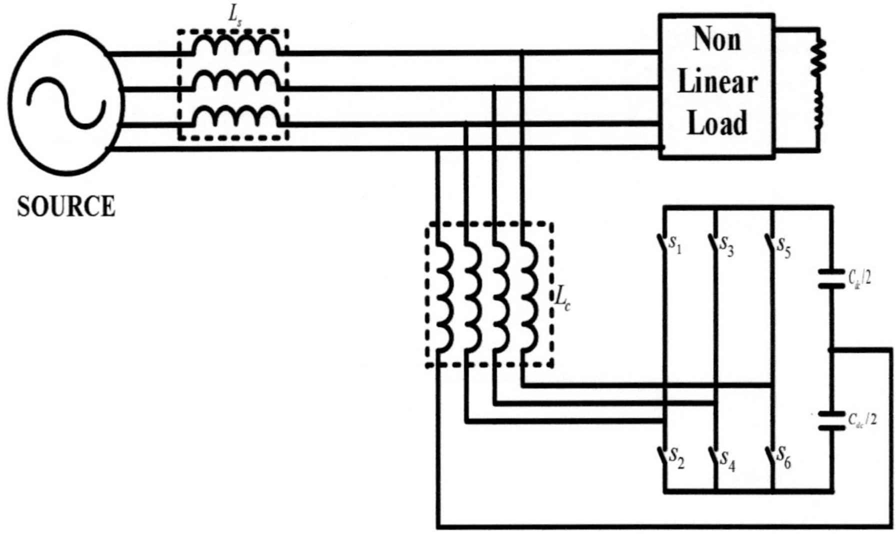

Figure 3.4 Three-phase four-wire shunt active filter.

is utilized in a variety of sectors to provide a comparable but anti-harmonic current in order to replace current harmonics.

3.4 Principle of the SAF System

In the 1970s, the fundamental idea of SAF was put forth. But in 1976, Guigi and Stricula put forth the SAF's initial concept.Compensation in current and reactive power is frequently referred to by the SAF. Figure 3.5 (a) displays the SAF's schematic diagram, while Figure 3.5 (b) displays the system's matching waveforms. By introducing anti-current harmonic components, SAF substitutes current harmonics. Injecting harmonic components produced by the load, it functions as a current source that is 180 degrees out of phase. The source current is therefore in phase with the sinusoidal voltage, and the SAF effect cancels out some of the harmonic currents at the load current [48]. Any kind of non-linear load that generates harmonics can be subject to this principle.

3.4.1 Characteristics of harmonics

Waveforms for voltage and current exhibit harmonic distortion. The majority of current distortion is produced by non-linear or electronic loads. Both

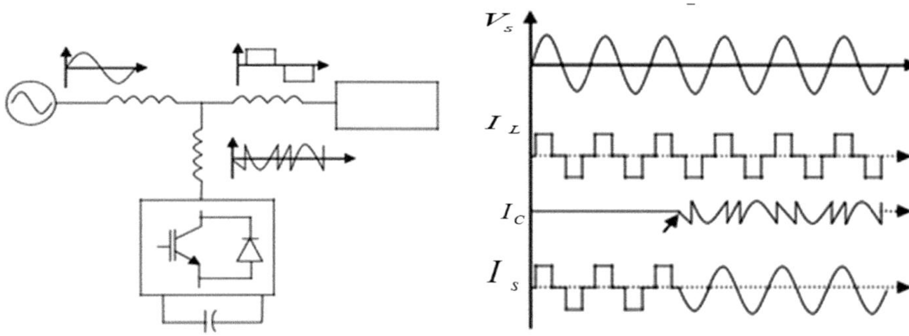

Figure 3.5 SAF working principle (a) schematic diagram of shunt SAF system and (b) schematic waveforms.

single- and three-phase non-linear loads are possible. Positive, negative, and zero sequence harmonics are produced by these loads. An efficient method for researching and analyzing harmonic distortion is the Fourier series. The purpose of the debate that follows is to obtain continuity and understanding from a research standpoint.

$$\begin{aligned} f(t) &= a_0 + \sum_{h=1}^{\infty} [a_h \cos(h\omega t) + b_h \sin(h\omega t)] \\ &= a_0 + \sum_{h=1}^{\infty} c_h \sin(h\omega t) + \psi_h, \end{aligned} \tag{3.1}$$

where $f(t)$ – periodic function of frequency 'f'
$\omega = 2\pi f$ – angular frequency
$T = \frac{1}{f} = \frac{2\pi}{\omega}$ – time period
c_h – hth harmonic of amplitude and ψ_h – phase.

The Fourier series coefficients are given by

$$a_0 = \frac{1}{T}\int_0^T f(t)dt = \frac{1}{2\pi}\int_0^{2\pi} f(t)dx, \text{ where } x = \omega t \tag{3.2}$$

$$a_h = \frac{2}{T}\int_0^T f(t)cosh(\omega t)dt = \frac{1}{\pi}\int_0^{2\pi} f(t)\cos(hx)dx \tag{3.3}$$

$$b_h = \frac{2}{T}\int_0^T f(t)\sin(\omega t)dt = \frac{1}{\pi}\int_0^{2\pi} f(t)\sin(hx)dx \tag{3.4}$$

$$c_h = \sqrt{a_h{}^2 + b_h{}^2} \text{ and } \psi_\text{h} = \tan^{-1}\left(\frac{\text{a}_\text{h}}{\text{b}_\text{h}}\right). \tag{3.5}$$

The following is the expression for the non-sinusoidal voltage or current waveform stretched into a Fourier series:

$$I\left(t\right) = \sum_{h=1}^{8} I_h \cos(\ h\omega t + \phi_h) \tag{3.6}$$

$$V\left(t\right) = \sum_{h=1}^{8} V_h \cos(\ h\omega t + \theta_h). \tag{3.7}$$

Here

$I_h = h\text{th}$ – peak current in harmonics, ϕ_h is the hth phase of harmonic current

$V_h = h\text{th}$ – peak voltage at harmonics, θ_h is a phase of harmonic voltage.

$\omega-$ angular frequency; $\omega = 2\pi f$, f is the power frequency.

3.4.2 Characteristics of harmonic compensation

From the grid, the instantaneous grid voltage V_s is written as

$$V_s\left(t\right) = V_{sm} \sin \omega t. \tag{3.8}$$

The source current at any given time I_s appears as

$$I_s\left(t\right) = I_L\left(t\right) - I_c\left(t\right). \tag{3.9}$$

The non-linear load current I_L comprises the current's fundamental and harmonic components, which are shown as

$$\begin{aligned} I_L\left(t\right) &= \sum_{h=1}^{\infty} I_h \sin(\ n\omega t + \varphi_h) \\ &= I_1 \sin(\omega t + \varphi_1) + \left(\sum_{h=2}^{\infty} I_h \sin(n\omega t + \varphi_h)\right). \end{aligned} \tag{3.10}$$

The eqn (3.10) has got three terms

$$I_L(t) = I_1 \sin(\omega t) + I_1 \cos(\omega t) + \sum_{h=2}^{\infty} I_h \sin(n\omega t + \varphi_h). \tag{3.11}$$

Using the grid voltage and the load current, the instantaneous load power is calculated. The computation of the load power is presented as

$$\begin{aligned} P_L(t) &= I_L(t) * V_s(t) \\ &= \left[V_{sm} I_1 \sin^2 \omega t * \cos\phi_1\right] + V_{sm} I_1 \sin\omega t * \cos\omega t * \sin\phi_1 \\ &\quad + V_{sm} \sin\omega\, t * \left(\sum_{h=2}^{\infty} I_h \sin(n\omega t + \varphi_h) \right) \\ &= P_f(t) + P_r(t) + P_h(t). \end{aligned} \tag{3.12}$$

From eqn (3.12), the real power is put in from the total load power and it is given as eqn (3.13)

$$P_f(t) = V_{sm} I_1 \sin^2 \omega t * \cos\phi_1 = V_s(t) * I_s(t). \tag{3.13}$$

If the entire reactive and harmonic power is supplied by the SAF, the source current $I_s(t)$ will be sinusoidal and in phase with the voltage. After harmonic mitigation, the source currents can be represented using eqn (3.14)

$$I_s(t) = P_f(t) / V_s(t) = I_1 \cos\phi_1 \sin\omega\ t = I_{sm} \sin\omega\, t, \tag{3.14}$$

where, $I_{sm} = I_1 \cos\phi_1 \sin\omega\ t$ – peak value of the source current.

There will be some switching losses in the VSI, so the grid needs to supply a small amount of active power to account for capacitor leakage and inverter switching losses, in addition to the real power delivered to the load. The total maximum current supplied by the source is given by eqn (3.15)

$$I_{sp} = I_{sm} + I_{sl}, \tag{3.15}$$

where,

$I_{sp} = I_1 \cos\phi_1 + I_{sl}$ – maximum value of the reference current that was captured,

I_{sl} – switching loss current.

If the SAF compensates the total reactive and distorted power; $I_s(t)$ will be sinusoidal and in phase with the voltage. The SAF compensation current is shown in eqn (3.16)

$$I_c(t) = I_L(t) - I_s(t). \tag{3.16}$$

The fundamental component of the load current must therefore be approximated in order to provide perfect and instantaneous compensation of reactive and distorted currents.

3.5 Design of the SAF System

The SAF system's parameters can be designed using a variety of techniques. The general procedures for system specifications and the values that are obtained from them are summarized in this section. To create a successful SAF, the following elements are utilized:

a) Choosing the device
b) Choosing the reference capacitor voltage
c) Interface inductor selection
d) Option for DC-link capacitor
e) Control unit

The following presumptions underlie the design of these components:

- source voltage sinusoidal;
- for designing the interface inductor, the grid side line current harmonic distortion is supposed to be 5%;
- limited reactive power compensation of the SAF;
- PWM-VSI functions in the linear variation mode;
- a function of the highest harmonic order that makes up for the chosen switching frequency.

a) Device selection:
The following variables determine the voltage and current rating of the chosen device:

- Switching frequency;
- Maximum current flowing through the device;
- Maximum voltage stress on the device

Maximum current flow: The extreme line current is equal to the peak current flowing over the switch S_A when it is closed (and S_A is open), which is one leg of the inverter. Peak line current and per-phase current are identical [47, 48]. For every device, the peak current flow is identical.

Maximum voltage stress: In view of the same situation of S_A being closed and $S_{A'}$ being open, the peak voltage is choked by the switch $S_{A'}$ which is V_{dc}. Generally, the choked voltage is chosen somewhat higher than V_{dc} so as to avoid device catastrophe [48].

The switching frequency required: Switching frequency should be as high as feasible to compensate for reactive and harmonic currents overall. The use of the PWM frequency is usually tenfold to compensate for the harmonic frequency [48].

Considering that we simply wish to replace the 50 Hz component, 500 Hz is ideal for reactive power compliance. To compensate for the 13th harmonic, however, we must substitute 6.5 kHz for the 650 Hz component in harmonic compensation. But because the gadget has a greater switching frequency, the switching losses go up [48].

b) Reference capacitor voltage selection:
Linear modulation mode $0 = m_a = 1$ is assumed for the SAF. The amplitude modulation factor m_a is given by eqn (3.17)

$$m_a = \frac{V_m}{V_{dc}/2}, \tag{3.17}$$

where $V_m = \sqrt{2}V_{c1}$.

Therefore,

$$V_{dc} = 2\sqrt{2}V_{c1} \text{ for } \mathrm{m_a} = 1, \tag{3.18}$$

where
V_{c1} – fundamental components in the AC-side of PWM inverter.

As per the principle of compensation, the current I_{c1} which is a fundamental current injects reactive power to the loads by the SAF. From the Figure 3.6 it can be seen that, the grid current I_{s1} needs to be in sync with the grid voltage. V_s and the fundamental compensation current I_{c1} must be orthogonal to V_s .

$$V_{c1} = V_s + j\omega L_c I_{c1}. \tag{3.19}$$

From eqn (3.19), the choice of V_{c1} must be set as per the ability of the system. The allowable V_{c1} should be in the range

$$V_s = V_{c1} = 2V_s. \tag{3.20}$$

$V_{c1} > 2V_s$ is not compulsory, because the maximum capacity generated is less than the case $V_{c1} = 2V_s$. In addition to that the larger V_{c1} means

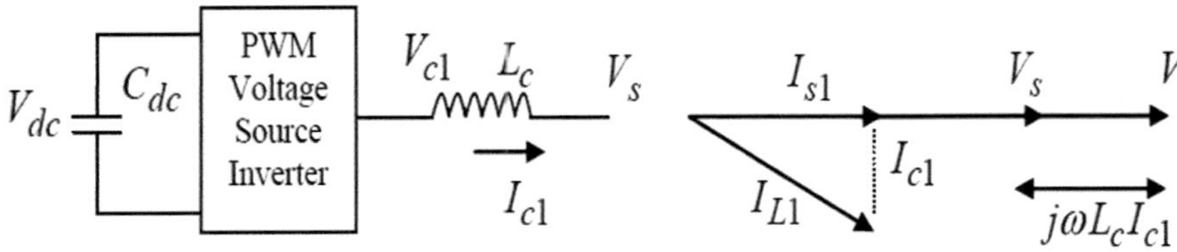

Figure 3.6 SAF vector diagram.

more V_{dc}, and thus more voltage stress may cause to the switches. Once V_{c1} is defined, the desired V_{dc} is calculated with the eqn (3.18) by putting $m_a = 1$.

But for typical non-linear load-like rectifiers, the DC voltage should adjust for the maximum harmonic order and is a function of the peak grid voltage [49]. Thus,

$$V_{dc} = 2\sqrt{2}V_{(ch)\max}, \tag{3.21}$$

where $V_{(ch)\max}$ is the peak voltage value of *Vc* (*t*) which includes harmonic mitigation as shown in eqn (3.22)

$$V_c\,(t) = Vs + \frac{6}{p}L_c\omega I_c\,(-\cos 5\,\omega t - \cos 7\,\omega t + \cos 1\,1\omega t + \ldots). \tag{3.22}$$

If there is a small inductor output filter because of a high switching frequency, the voltage V_{c1} is around equal to the grid voltage, therefore the eqn (3.18) will be

$$V_m = 2\sqrt{2}V_s. \tag{3.23}$$

Supposing a value of 1.1 V_{dc}, the DC voltage is controlled to be *10%* more than the maximum input voltage. Moreover, the desired V_{dc} is abridged. Third and ninth harmonics are modulated in PWM because of the sine reference. Equation (3.24) indicates the grid voltage in terms of the minimal necessary voltage.

$$V_{dc} = \frac{2\sqrt{2}}{1.155}V_s. \tag{3.24}$$

c) Interface inductor selection:

There must be enough airflow between the conductor and the inductor core for the interface inductor to have a low flow density. In order to estimate the interface inductor value, the following conditions can be applicable [49].

- Provide a way to compensate for currents (current change rate) for harmonics and reactive energy.
- Verify that the PWM-inverter's voltage source and output current ripple to a predetermined quality level.

First approach: Reactive power compensation and harmonic current mitigation are essential to the design of the interface inductor L_c. Additionally, compensated current waveforms from inverters can be filtered using an inductor. The maximum harmonic voltage of the sinusoidal PWM-VSI is supposed to be at the frequency (m_f) when it operates in linear modulation mode. Given this harmonic content, eqn (3.25) provides the SAF's ripple current.

$$I_{ch} = I_{ch}(m_f\omega) = \frac{V_{ch}(m_f\omega)}{L_c(m_f\omega)}. \tag{3.25}$$

In this case, the frequency modulation ratio of the PWM-VSI is represented by the subscript *h*, which stands for the harmonics m_f. The ripple attenuation factor (RAF) is defined as follows for the quantitative representation (3.26):

$$\mathrm{RAF} = \frac{I_{ch}}{I_{c1}, \mathrm{rated}} \tag{3.26}$$

The three-phase reactive power Q_{c1} delivered from the SAF to the system is calculated from the vector diagram in Figure 3.6 is used to compute the system's SAF.

$$Q_{c1} = 3V_s I_{c1} = 3V_s \frac{V_{c1}}{\omega L_c}\left(1 - \frac{V_s}{V_{c1}}\right). \tag{3.27}$$

By solving above two eqn (3.24) and (3.26) simultaneously the values of L_c and V_{c1} are calculated.

Second approach: The chosen criterion for the interface inductor's design is the peak ripple current. The ripple current is calculated under no-load conditions, with the assumption that inductor resistance has no effect. In this scenario, the inverter's reference voltage is the same as the grid voltage [49]. The necessary inductor is therefore provided by

$$L_c = \frac{V_s}{2\sqrt{6} f_s \Delta I_{c(p-p)\max}}, \tag{3.28}$$

where $\Delta I_{c(p-p)\max}$ is 15% peak compensation current.

Third approach: To compensate for a completely inductive load, the filter must grid the maximum current $I_{\max}$, which is known as the inductor value.

$$L_{c\min} = \frac{\Delta V}{\omega I_{\max}}, \tag{3.29}$$

where ΔV is the source voltage to inverter voltage difference, which is determined by the modulation index and DC voltage values;
ω is the value of the fundamental frequency.

This is the minimum value of the inductor that should be utilized to maintain the ripple current at a low level. In order to accommodate increased surge and harmonic currents, the inductor value is reduced in an active power line conditioner.

In this condition, other criteria can be evaluated. Imposing a fixed switching frequency f_s the maximum $\frac{di}{dt}$ of the current to be compensated, and the ripple current ΔI_c inductance value is given by

$$L_{c\max} = \frac{\Delta V_{\max}}{4f_s \Delta I_{c(p-p)}}. \tag{3.30}$$

Global filter performance won't be impacted by the inductor design if the ripple is decreased by raising the switching frequency or managing dynamics. Additionally, the connecting inductances separate the ac source voltage from the output inverter voltage. When all of these parameters are taken into account, the inductor value between the minimum and maximum current extremes is regarded as

$$\frac{\Delta V}{\omega I\text{max}} = Lc = \frac{\Delta V\text{max}}{4f_s \Delta I_{c(p-p)}}. \tag{3.31}$$

Fourth approach: The active power line conditioner's maximum ripple current constraint determines the inductor needed for current regulation. The fundamental frequency component of the output voltage's zero-crossings in a six-switching power transistor experiences the highest ripple current [55]; the inductor value is calculated as

$$L_c = \frac{V_{dc}}{6f_c \Delta I_{(p-p)\max}}. \tag{3.32}$$

In many SAF applications, the first method is frequently employed to construct the interface inductor. The harmonics produced by the power transistors' switching activity are suppressed by the interface inductor. The voltage source inverter and high-frequency components are isolated and smoothed by the inducer. The inverter's switching frequency and the available driving voltage across the interface inductor set limits on the control of the current wave shape.

d) DC-link capacitor selection:

The power source's DC-link capacitor serves two crucial purposes: (i) In the solid state, it nearly always maintains a DC voltage with slight oscillations. (ii) During the transient, it serves as a power reserve to supply a genuine power differential between the source and the load. In order to minimize the loss to the SAF, the source's actual power must, in the worst scenario, match the actual power needed for that load with less power. However, the energy balance between the mains is crucial, and as a result, the load will be disrupted when the load condition changes. The DC-link capacitor must make up for this discrepancy in actual power. The current reference value must be linearly changed to convert the actual power output to the source in order for the power line operating condition to operate as efficiently as possible. The capacitor charges and releases this actual power, which includes the power that the load actually uses. To support the rise or fall in the voltage applied by the linear or non-linear line, the DC-link capacitor's stored energy is either instantly released or determined [49].

First approach: On the basis of instantaneous power flow on both the DC and AC sides of the PWM-inverter, the DC-link capacitor was designed. The selection of C_{dc} is governed by reducing the voltage ripple. The DC-link capacitor C_{dc} is found from rated active power line conditioner current $I_{c1,\text{rated}}$ and peak-to-peak voltage ripple $V_{dr,(p-p)\max}$ [49]. It is defined as

$$C_{dc} = \frac{\pi * I_{c1,\text{rated}}}{\sqrt{3}\omega V_{dr,(p-p)\max}}. \tag{3.33}$$

Second approach: During the transitory state, the capacitor must grid the load's actual power demand for one grid voltage cycle [49]. Therefore, the following is the capacitor value equation based on the principle:

$$C_{dc} = \frac{2E_{\max}}{{V_{dc}}^2 - V_{dc}\text{min}^2}, \tag{3.34}$$

where $2E_{\max}$ – maximum energy that the capacitor can supply to the grid while it is in a transient state.

Third approach: The negative-sequence load current causes ripple power at 2ω to be handled by the DC-link capacitor. The value of the minimum capacitor is determined as

$$C_{dc} = \frac{S}{2\omega V_{dc}\Delta V_{dc}}. \tag{3.35}$$

Fourth approach: The measurement of the lowest rank harmonic current, I_h, is the foundation of the first method. The value of the capacitor is established by

$$C_{dc} = \frac{I_h}{eV_{dc}\omega}. \tag{3.36}$$

Calculating the energy supplied by the active power line conditioner during a half-cycle period of the power pulsation associated with the first two harmonics–the fifth and seventh for a three-phase rectifier–is the foundation of the second method. Selecting a suitable rate ripple (example $e = 5\%$) the capacitor C_{dc} is calculated as follows

$$C_{dc} = \frac{V_s\sqrt{{I_5}^2 + {I_7}^2 - 2I_5I_7cos(\ 5a - 7a)}}{2\omega {V_{dc}}^2 e}, \tag{3.37}$$

where α – firing angle delay.

e) Control unit: The PWM-current control methodology and the reference current extraction method make up the control unit.

3.6 Reference Current Extraction Techniques

The literature discusses a number of control strategies for separating the reference currents from the contaminated load currents. Unit vector control theory, Fryze power theory, instantaneous real-reactive power (p-q) theory, modified p-q theory, SRF theory, and modified-SRF theory are a few of the reference current extraction techniques. The Apq-SRF methodology, a unique reference current extraction method, has been presented.

3.6.1 Unit vector control with PI/type-1 fuzzy/type-2 fuzzy DC-link voltage controller

SAF controlling includes reference current extraction technique, pulse generation technique, and DC-link voltage controller. The desired current reference is taken out from the polluted line-current by means of the unit vector control beside with the PI/type-1 fuzzy/type-2 fuzzy controller as shown in Figure 3.7.

Unit sine vector

The instantaneous grid voltages are measured and displayed in eqn (3.38) to create the unit sine vector.

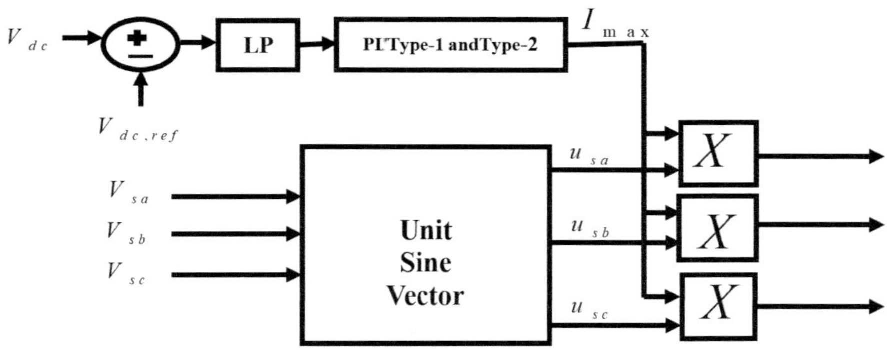

Figure 3.7 Block diagram of the unit vector control theory.

$$
\begin{aligned}
V_{sa} &= V_{sm}\sin(\omega t) \\
V_{sb} &= V_{sm}\sin(\omega t - 120^\circ) \\
V_{sc} &= V_{sm}\sin(\omega t + 120^\circ),
\end{aligned} \tag{3.38}
$$

where V_{sm} – maximum value of the grid voltage,
$\omega = 2pf$ – fundamental angular frequency.

To achieve the harmonic mitigation, three-phase supply currents and unity power factor are estimated using the unit sine vector templates. The grid voltages and their maximum value are in phase with the unit sine vector templates [50]. Equation (3.39) provides the unit sine vector templates.

$$
\begin{aligned}
u_{sa} &= V_{sa}/V_{sm} = \sin\omega t \\
u_{sb} &= V_{sb}/V_{sm} = \sin(wt - 120^\circ) \\
u_{sc} &= V_{sc}/V_{sm} = \sin(\omega t + 120^\circ).
\end{aligned} \tag{3.39}
$$

The unit sine vector template's amplitude is unity in steady state. The unit sine vector's value under the dynamic state is contingent upon variations in the load. The necessary reference currents are created by multiplying the unit sine vector by the maximum amplitude of the anticipated reference current.

3.6.2 Reference current generation

The PI/type-1 fuzzy/type-2 fuzzy controlling is used to control the DC-link capacitor voltage and forecast the reference current's peak value. The PI/type-1 fuzzy/type-2 fuzzy controlling's output is regarded as the reference current's or source's peak value. The predicted reference current as given in eqn (3.40) is having of two components: (i) power frequency real power component of the load current, (ii) power loss component of the SAF including switching losses. The block diagram of the unit vector control to predict the desired reference currents is as shown in Figure 3.7.

To determine the maximum current, the DC capacitor voltage is measured and compared to the reference voltage. The voltage error $e\,(v) = V_{dc-\text{ref}} - V_{dc}$ and the change of error are used as input to the PI/type-1 fuzzy/type-2 fuzzy controller.

$$\begin{aligned} i_{ca}* &= I_{\max}u_{sa} = I_{\max}\sin\omega t \\ i_{cb}* &= I_{\max}u_{sb} = I_{\max}\sin(\omega t - 120°) \\ i_{cc}* &= I_{\max}u_{sc} = I_{\max}\sin(\omega t + 120°). \end{aligned} \tag{3.40}$$

3.6.3 PI DC-link voltage controller

The DC capacitor voltage is controlled and the necessary amplitude of the current reference is measured using the conventional PI controller. It is anticipated that the source will only supply the basic power frequency component of the current load whenever it supplies a non-linear or inductive load, with adjustment providing the active/active component. Furthermore, compensation offers anti-distorted power, which only appears in the energy's active component [51]. The DC-link capacitor voltage increases (or decreases) in response to changes in load during transient situations when the capacitor draws (or loses) excess capacity (or shortfall). The current reference to this treatment is seen in this preservation of the philosophy of power. A non-linear load's distortion in the intermediate voltage is linked to the divergence from the capacitor voltage. The output power is employed to lower and sustain the capacitor voltage with the aid of this substance.

The PI control algorithm is calculated based on the DC power of the DC-link, the three-phase supply voltages, and the currents (ignoring the download load and compensation muscles). VSI switch pumps are generated using the indirect PWM-current process that reduces the sensors and controller complexity.

The purpose of the low-pass filter (LPF) is to filter overlapping components and high-order handles while permitting fundamental components. Through the LPF, the voltage error signal travels above the primary signal. The definition of the transfer function's PI controller is provided by

Table 3.1 Ziegler-Nichols approximation method gain calculations.

Control type	K_p	K_i	K_d
P	0.50 K_u	–	–
PI	0.45 K_u	1.2 K_p/P_u	–
PID	0.60 K_u	2 K_p/P_u	$K_pP_u/8$

eqn (3.41).

$$H(s) = K_P + K_I/s. \tag{3.41}$$

Either heuristically or mathematically, the proportional and integral gains are chosen. The K_P determines the transient response of the DC capacitor voltage. It is calculated as $K_P = 2\zeta_{\omega nv}C$, where, $\zeta = \sqrt{2}/2$-damping ratio, C – DC capacitor, ω_{nv} – natural frequency, chosen as the supply power frequency. Similarly, the integral gain is calculated by $K_I = C{\omega_{nv}}^2$ that gives settling time and removes the DC capacitor voltage's steady state error.

Many tuning methods like mathematical modeling, tuning with intelligent controllers (fuzzy, neural, and PSO) are available in the literature. These methods are complex and difficult to tune the gain values. Therefore, the Ziegler-Nichols approximation approach is employed in this study to determine the gain values [51]. First K_i and K_d gains are first set to zero. The K_p is increased until it reaches the gain K_u, where the output of the loop starts to oscillate. K_u and the oscillation period P_u are used to set the gains as shown in Table 3.1. These gains are applicable to the ideal, parallel form of the PID controller. For the standard PID controllers, the time constants for integral and derivative T_i and T_d are purely dependent on the oscillation period P_u.

This controller anticipates the magnitude of maximum reference current $I_{\max}$ by controlling the DC capacitor. When the expected maximum current is multiplied by the unit sine vector output, the necessary reference currents are obtained.

VSI switching pulses are produced by an indirect PWM-current controller by comparing the reference current to the measured real current.

3.6.4 Type-1 fuzzy logic DC-link voltage controller

The fuzzy set concept, first proposed by Zadeh in 1965, is the foundation of fuzzy logical control. The fuzzy logic concept involves a transition between membership and non-membership functions. Therefore, the bounds of a fuzzy set are helpful for creating structure even when they are not clear and complex. When compared to the intended reference value, the DC-link capacitor voltage is appropriate for applying the robust control algorithm of

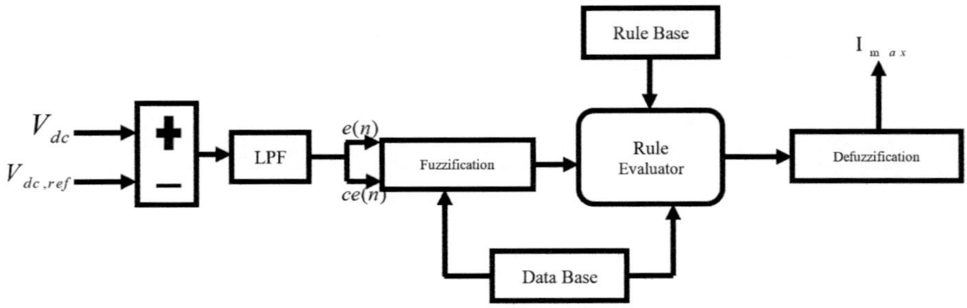

Figure 3.8 Block diagram of the type-1fuzzy logic controller.

the SAF algorithm in a closed loop. The error signal $e\,(v) = V_{dc-\text{ref}} - V_{dc}$ passes through a Butterworth low-pass filter that allows only the fundamental component [52].

The error signal $e\,(n)$ and change of error signal $ce\,(n)$ are utilized as inputs for fuzzy processing as shown in Figure 3.8. The fuzzy logic controller output anticipates the value of maximum reference current $I_{\max}$.

The following describes the fuzzy logic controller:

Seven fuzzy sets of the following have been used:

- NB, NM, NS, ZE, PS, PM, and PB for each input and output variable;
- triangle membership function;
- Mamdani-type min-operator for implication; and centroid method for defuzzification.

Fuzzification: Fuzzy logic utilizes linguistic variables in place of numerical variables. The error $e(n)$, change of error $ce(n)$, and output of maximum reference current I_{max} are selected as membership functions. Figure 3.9 shows the labels for these: negative large (NB), negative medium (NM), negative tiny (NS), zero (ZE), positive small (PS), positive medium (PM), and positive big (PB). Fuzzification is capable of translating a numerical variable into a verbal one [52].

Rule elevator: The fuzzy logic rules are linguistic variables in nature and a typical rule can be written asR_k: If $e(n)$ is *Ai* and *ce*(*n*) is *Bi* then $I_{\max}$is *Ci* where, *Ai*, *Bi*, and *Ci* are linguistic variables of $e(n)$, *ce*(*n*), and $I_{\max}$ and it represents the degree of membership function.

Defuzzification: A linguistic variable has a series of changed control outputs that are produced using fuzzy logic techniques. Defuzzification is the process of turning linguistic variables into a clear value appropriate for real-world

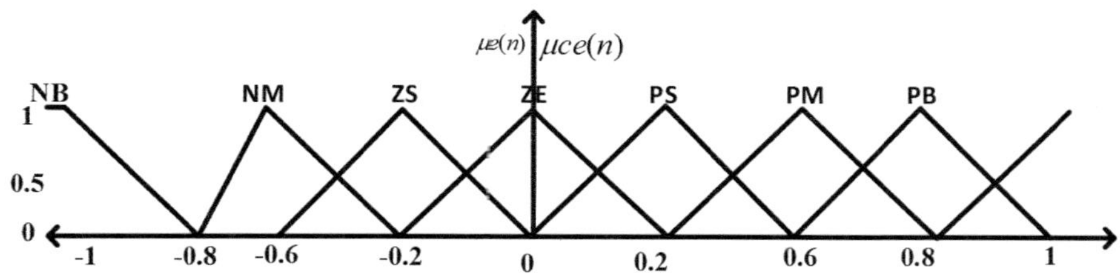

(a) Input variables "*e (n)* and *ce(n)*."

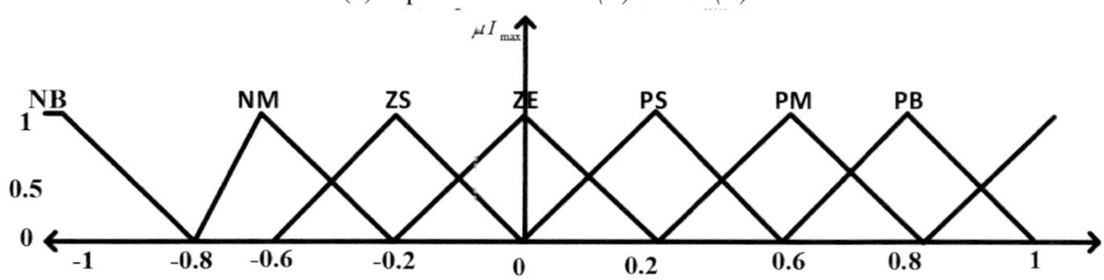

(b) Output variable "$I_{\max}$."

Figure 3.9 Membership functions (a) the input variables $e(n)$, ce(n), and (b) output variable $I_{\max}$.

uses. Among the defuzzification methods are centroid, big of maximum (LOM), smallest of maximum (SOM), middle of maximum (MOM), and bisector. The centroid (or center of gravity) method is used in this proposed research for simplicity and accuracy. Crisp output is obtained by using $output = \sum A_i * x_i / \sum A_i$.

Database: The fuzzifier and defuzzifier will use the membership function that is stored in the database. The processor and accessible memory of the digital controller chip are negotiating.

Rule base: Linguistic control rules desired by the rule evaluator stored by the rule base. The 49-rules used in this research are presented in Table 3.2.

Table 3.2 Rule base table with 49-rules.

e(n) \ ce(n)	NB	NM	NS	ZE	PS	PM	PB
NB	NB	NB	NB	NB	NM	NS	ZE
NM	NB	NB	NB	NM	NS	ZE	PS
NS	NB	NB	NM	NS	ZE	PS	PM
ZE	NB	NM	NS	ZE	PS	PM	PB
PS	NM	NS	ZE	PS	PM	PB	PB
PM	NS	ZE	PS	PM	PB	PB	PB
PB	ZE	PS	PM	PB	PB	PB	PB

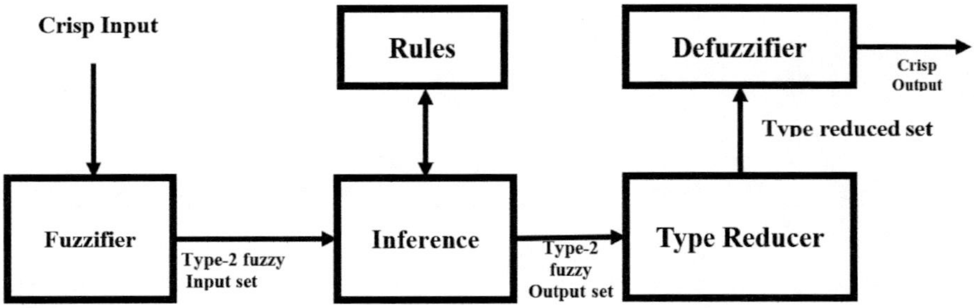

Figure 3.10 Type-2 fuzzy logic block diagram.

The fuzzy controller predicts the value of peak reference current $I_{\max}$. The predicted value of peak-current multiplied with unit sine vector determines the reference currents.

3.6.5 Type-2 fuzzy logic DC-link voltage controller

Type-2 fuzzy logic handles uncertainties in a better manner as compared with type-1. It is having a third dimension called foot point of uncertainty (FOU) allows better control than type-1 fuzzy logic [52]. The block diagram of the type-2 fuzzy logic for regulating the DC-link voltage is shown in Figure 3.10.

The linguistic rules which have been used for type-1 fuzzy set is also used for type-2 fuzzy set.

Remarks: Unit vector is very simple to implement for extracting the SAF reference currents. The design of PI control is difficult, it requires precise mathematical equations. Nowadays, the controllers based on fuzzy logic of either type-1 or type-2 are used for SAF applications. The merits of fuzzy logic over the conventional controllers are: (i) accurate numerical calculations are not required, (ii) can be used with imprecise inputs, (iii) works well for non-linear systems, and (iv) efficient than PI controllers.

3.7 Fryze Power Theory

Similar to the initial load current, this approach extracts the three-phase average real power from the source by presenting a minimum RMS value. It guarantees that the grid voltage and load current are linear [53]. In Figure 3.11, the generalized Fryze current control approach is displayed.

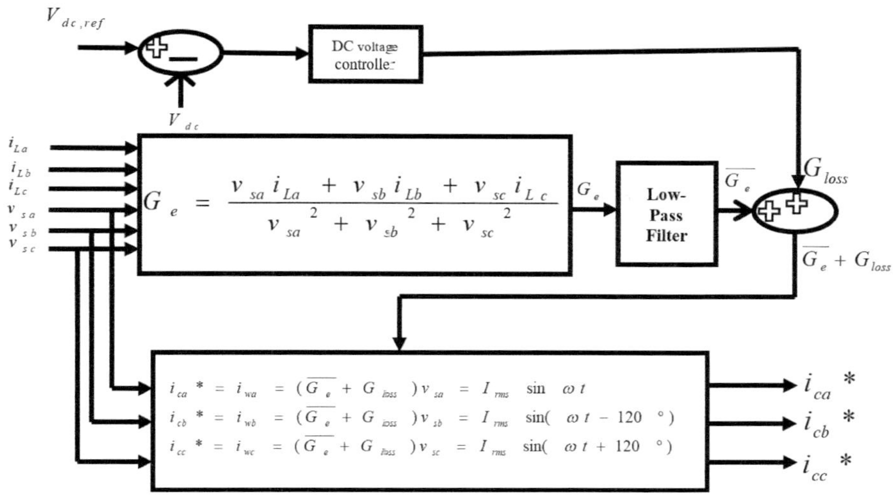

Figure 3.11 Block diagram of the generalized Fryze power theory algorithm.

The instantaneous conductance (G_e) is calculated from the three-phase instantaneous real power ($P3\phi$) as given in eqn (3.42)

$$\begin{aligned} p_{3\phi} &= v_{sa}(t)\, i_{sa}(t) + v_{sb}(t)\, i_{sb}(t) + v_{sc}(t)\, i_{sc}(t) \\ &= p_{sa}(t) + p_{sb}(t) + p_{sc}(t)\,. \end{aligned} \tag{3.42}$$

The instantaneous value of the grid phase voltages is used to compute the root mean square voltage, which is provided as eqn (3.43)

$$V = \sqrt{{v_{sa}}^2(t) + {v_{sb}}^2(t) + {v_{sc}}^2(t)}. \tag{3.43}$$

Using three-phase instantaneous phase voltages v_{sa} ,v_{sb}, v_{sc} and load currents i_{sa}, i_{sb}, and i_{sc}, the *instantaneous* conductance (Ge) is calculated and is shown in eqn (3.44)

$$G_e = \frac{v_{sa} i_{La} + v_{sb} i_{Lb} + v_{sc} i_{Lc}}{{v_{sa}}^2 + {v_{sb}}^2 + {v_{sc}}^2}. \tag{3.44}$$

In order to calculate the average conductance ($\overline{G_e}$), the conductance (G_e) passes through the Butterworth design-based LPF. Since the LPF cut off frequency is 50 Hz it only allows the power frequency component for the real current calculation. Multiplying $\overline{G_e} + G_{\text{loss}}$ by the phase voltages, the instantaneous real components of the load currents are directly obtained.

This control strategy adheres to the conventional current direction by assuming that the SAF draws the inverse of the non-active current from the load.

In order to compensate for the switching and conducting loss in the SAF, which causes the DC capacitor voltage to deplete, a tiny amount of real power is extracted by adding an additional real component of current. This is done by adding the G_{loss} and $\overline{G_e}$. Equation (3.45), which is calculated from the actual currents, represents the necessary reference grid currents.

$$\begin{aligned} i_{ca}* &= i_{wa} = \left(\overline{G_e} + G_{\text{loss}}\right) v_{sa} = I_{\text{rms}} \text{sin}\omega t \\ i_{cb}* &= i_{wb} = (\overline{G_e} + G_{\text{loss}}) v_{sb} = I_{\text{rms}} \text{sin}(\omega t - 120°) \\ i_{cc}* &= i_{wc} = (\overline{G_e} + G_{\text{loss}}) v_{sc} = I_{\text{rms}} \text{sin}(\omega t + 120°), \end{aligned} \tag{3.45}$$

where I_{rms} – rms line current.

The comparison of reference current and actual SAF compensating current generates switching pulses.

Remarks: Because it works directly with the traditional *abc*-phase voltage and line currents, this method has the advantage of lowering the reference current computation. This control is easy to build because no Clarke transformation is required. However, this hypothesis is unrelated to the three-phase instantaneous reactive power for compensating for the loads' reactive power need.

3.8 Instantaneous Real and Reactive Power (p-q) Theory

H. Akagi and associates first suggested the instantaneous active and reactive power control approach (p-q control strategy) in 1984. Since then, it has undergone numerous revisions and interpretations. Based on Clarke's ("α-β") translation of voltage and current signals to create compensating signals, the instantaneous active and reactive power (p-q) control approach has been employed extensively. Figure 3.12 shows the full reference current generation utilizing a PI controller and a traditional p-q control scheme [54]. The active power component is controlled by a PI/type-1 fuzzy/type-2 fuzzy controller to keep the DC-link voltage constant.

This tiny amount of active current is managed by the PI/type-1 fuzzy/type-2 fuzzy controller, and the current controller then adjusts this current to keep the DC-link capacitor voltage steady. To do this, the control circuit detects and compares the DC-link voltage with the reference

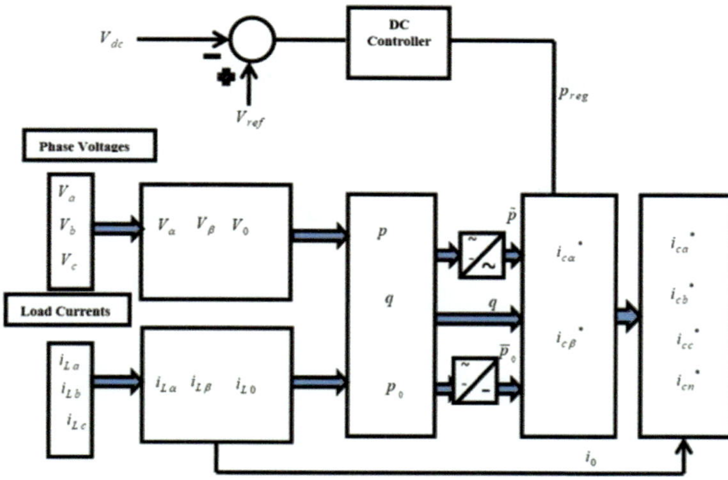

Figure 3.12 p-q control strategy for reference current extraction.

voltage setting. The PI/type-1 fuzzy/type-2 fuzzy controller then receives the difference.

The amount of active current required to maintain the DC-link voltage is determined by the PI/type-1 fuzzy/type-2 fuzzy controller based on the voltage differential. An active current, or the corresponding power flow required to maintain the DC-link voltage, is the output of the PI/type-1 fuzzy/type-2 fuzzy controller. It is a component of the current controller's reference current, which regulates the inverter to supply the necessary compensation current.

It is possible to calculate the instantaneous active and reactive power using the converted voltage and current information from Figure 3.13. Using low-pass and high-pass filters, harmonic active and reactive powers are derived from instantaneous active and reactive powers. Using the inverse "α-β" transformation, compensating directives in terms of either currents or voltages are produced from harmonic active and reactive powers. The nonlinear load's instantaneous active and reactive powers, p and q, are used to generate currents in the SAF.

Transformation of the phase voltages V_a, V_b, and V_c and the load currents i_{La}, i_{Lb}, and i_{Lc} into the $\alpha - \beta$ orthogonal coordinates using Clarke's transformation which is as shown in Figure 3.13. The compensation objectives of active power filters are the harmonics present in the input currents. The phase voltages and line currents in $\alpha - \beta - 0$ co-ordinates is as shown in eqn (3.46)

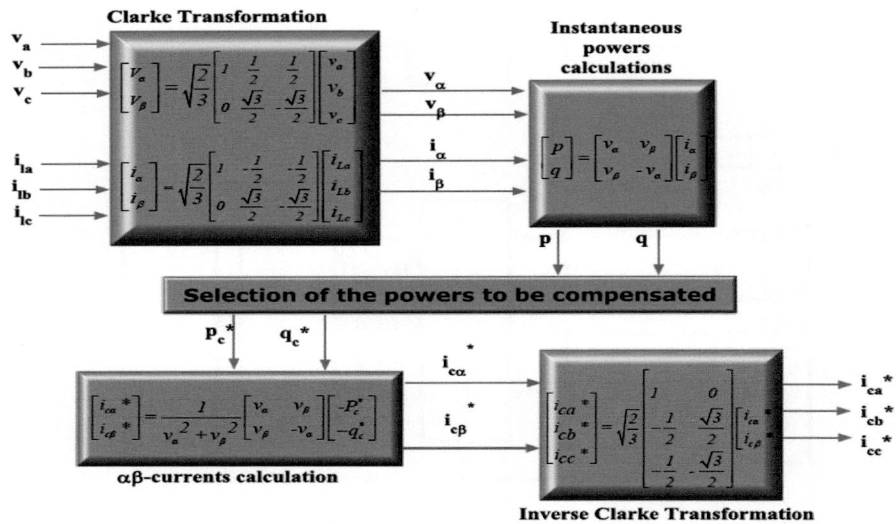

Figure 3.13 p-q control strategy calculations.

and (3.47)

$$\begin{bmatrix} V_0 \\ V_\alpha \\ V_\beta \end{bmatrix} = \sqrt{\frac{2}{3}} \begin{bmatrix} \frac{1}{\sqrt{2}} & \frac{1}{\sqrt{2}} & \frac{1}{\sqrt{2}} \\ 1 & -\frac{1}{2} & -\frac{1}{2} \\ 0 & \frac{\sqrt{3}}{2} & -\frac{\sqrt{3}}{2} \end{bmatrix} \cdot \begin{bmatrix} V_a \\ V_b \\ V_c \end{bmatrix} \tag{3.46}$$

$$\begin{bmatrix} i_0 \\ i_\alpha \\ i_\beta \end{bmatrix} = \sqrt{\frac{2}{3}} \begin{bmatrix} \frac{1}{\sqrt{2}} & \frac{1}{\sqrt{2}} & \frac{1}{\sqrt{2}} \\ 1 & -\frac{1}{2} & -\frac{1}{2} \\ 0 & \frac{\sqrt{3}}{2} & -\frac{\sqrt{3}}{2} \end{bmatrix} \cdot \begin{bmatrix} i_{La} \\ i_{Lb} \\ i_{Lc} \end{bmatrix}. \tag{3.47}$$

Equation (3.48) provides the instantaneous real power in zero sequence.

$$p_0 = V_0 i_0. \tag{3.48}$$

Equation (3.49), in $\alpha - \beta$ coordinates, provides the instantaneous real power.

$$p = V_\alpha i_\alpha + V_\beta i_\beta. \tag{3.49}$$

In $\alpha - \beta$ coordinates, the instantaneous imaginary power is provided by eqn (3.50).

$$q = V_\alpha i_\beta - V_\beta i_\alpha. \tag{3.50}$$

In the form of $\alpha - \beta$ coordinates of voltages and currents, p and q are provided by eqn (3.51).

$$\begin{bmatrix} p \\ q \end{bmatrix} = \begin{pmatrix} V_\alpha & V_\beta \\ -V_\beta & V_\alpha \end{pmatrix} \cdot \begin{bmatrix} i_\alpha \\ i_\beta \end{bmatrix}. \tag{3.51}$$

The reference currents interms of $\alpha - \beta$ coordinates are given by eqn (3.52)

$$\begin{bmatrix} I_\alpha{}^* \\ I_\beta{}^* \end{bmatrix} = \frac{1}{V_\alpha{}^2 + V_\beta{}^2} \begin{bmatrix} V_\alpha & -V_\beta \\ V_\beta & V_\alpha \end{bmatrix} \begin{bmatrix} p^\sim - \overline{p_0} \\ q \end{bmatrix}, \tag{3.52}$$

where

$$I_0{}^* = I_0 \tag{3.53}$$

$$\begin{bmatrix} I_{ca}{}^* \\ I_{cb}{}^* \\ I_{cc}{}^* \end{bmatrix} = [C]^{-1} \begin{bmatrix} I_0 \\ I_\alpha{}^* \\ I_\beta{}^* \end{bmatrix}, \tag{3.54}$$

$$\text{where } [\mathrm{C}] = \sqrt{\frac{2}{3}} \begin{bmatrix} \frac{1}{\sqrt{2}} & \frac{1}{\sqrt{2}} & \frac{1}{\sqrt{2}} \\ 1 & -\frac{1}{2} & -\frac{1}{2} \\ 0 & \frac{\sqrt{3}}{2} & -\frac{\sqrt{3}}{2} \end{bmatrix}. \tag{3.55}$$

The power flow diagram under non-linear loading conditions is as shown in Figure 3.14

where

$\overline{p_0}$ = average amount of the instantaneous zero-sequence real power, or the energy arriving at the load per unit of time, from the source via the $V_0 \& i_0$.

$p_0{}^\sim$ = altered quantity of the instantaneous zero-sequence real power which indicates that the energy per unit time that is transferred among the source and the load via the zero-sequence values. Three-phase systems with a single neutral wire can use zero-sequence power. Systems must include unbalanced voltages and currents, as well as third harmonics in both voltage and current of at least one phase, in order to have zero-sequence components.

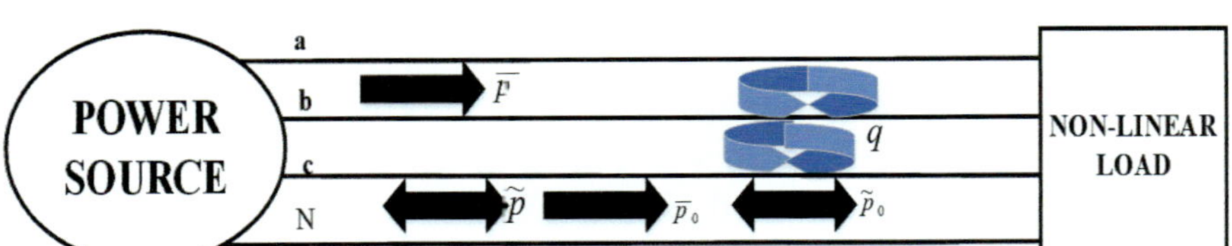

Figure 3.14 Power flow diagram for p-q control.

$\overline{p}$ = average amount of the instantaneous real power, or the energy per unit of time, that is supplied to the load via the *a-b-c* coordinates in a balanced manner, that is, only in a positive sequence.

$p^{\sim}$ = altered amount of the instantaneous real power, or the energy per unit of time that circulates via the *a-b-c* coordinates between the source and the load.

q = The power that is transferred between the load's lines (L-L) is known as instantaneous reactive power.Although there is no energy exchange or circulation between the source and the load, this component indicates that undesired currents are available to flow between the phases. Regarding a balanced sinusoidal power source and a balanced load, with or without harmonics, the conventional reactive power is equal to q, the mean value of the instantaneous imaginary power, whether or not harmonics are present.

One of the greatest techniques for controlling shunt active filters is the instantaneous theory. P-Q theory's key characteristics are as follows:

- It has a three-phase theory built in;
- It operates on instantaneous powers, provides a good steady state and dynamic response;
- It can be applied to variable non-linear loads; and
- The computations are very easy (just algebraic equations must be used with standard processors).

Remarks: Though the instantaneous real and reactive power (p-q) theory is applicable for balanced, unbalanced, and dynamic conditions but this theory does not work well for distorted voltage supply conditions. To make it work even for distorted supply conditions a modified p-q control theory i.e., sinusoidal control strategy can be applied to extract fundamental supply voltages and separate distorted supply voltage components.

3.9 Modified p-q Control Strategy

The three-phase voltage signal supplied by the power line is converted into the positive sequence fundamental signal by the positive sequence voltage detector, as seen in Figure 3.15. The PLL control circuit is a crucial component of the positive sequence voltage detector since it monitors the positive sequence voltage at the fundamental frequency of extremely distorted and unbalanced three-phase signals. The measured signals' phase angle and fundamental frequency, which may be imbalanced and contain harmonics, are precisely determined by the synchronization circuit [54].

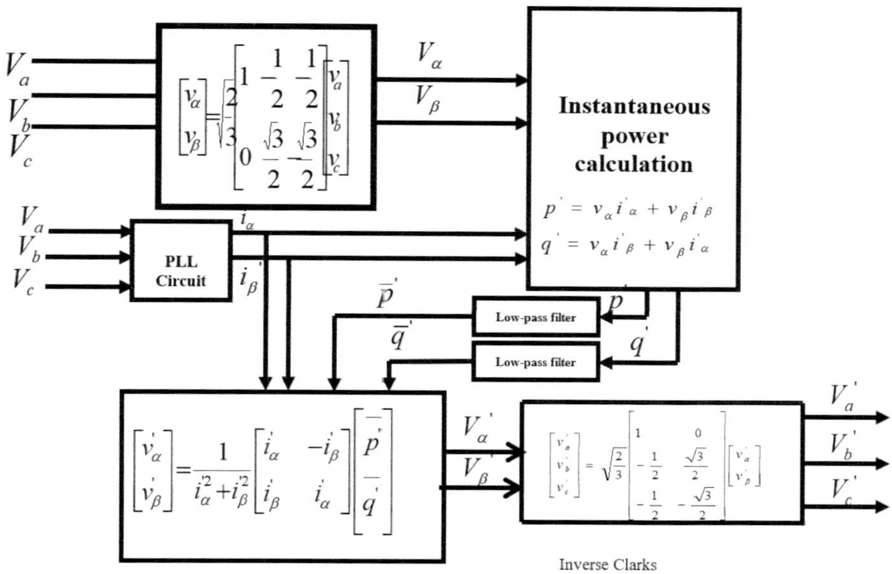

Figure 3.15 Fundamental positive sequence voltage detector.

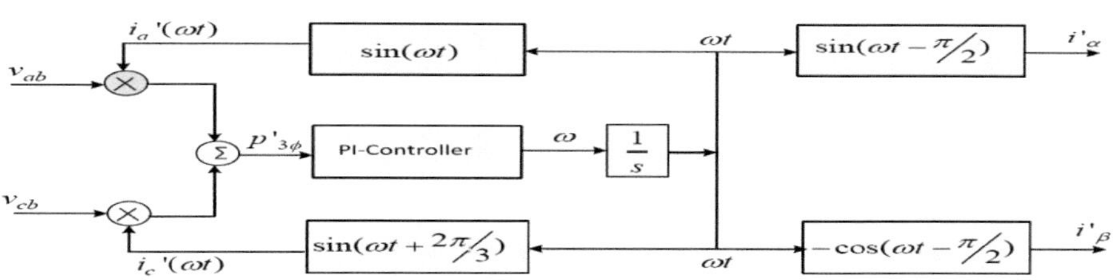

Figure 3.16 Functional block diagram of the PLL circuit.

The fundamental positive-sequence component of a three-phase generic input signal's system frequency and phase angle are automatically determined by the PLL circuit, as seen in Figure 3.16. Even when system voltages are severely distorted, modified p-q theory works incredibly well. Equation (3.56) provides the algorithm's fake instantaneous active power expression.

$$p'_{3\varphi} = v_a i'_a + v_b i'_b + v_c i'_c = v_{ab} i'_a + v_{cb} i'_c. \tag{3.56}$$

Eqn (3.57) is considered for $i'_a + i'_b + i'_c = 0$. In actual, $p'_{3\varphi}$ is not related to any instantaneous active power of the power system, although it could be considered as a variable in the PLL circuit with a dimension of power. The fictitious current feedback signals $i'_a(\omega t) = \sin(\omega t)$ and

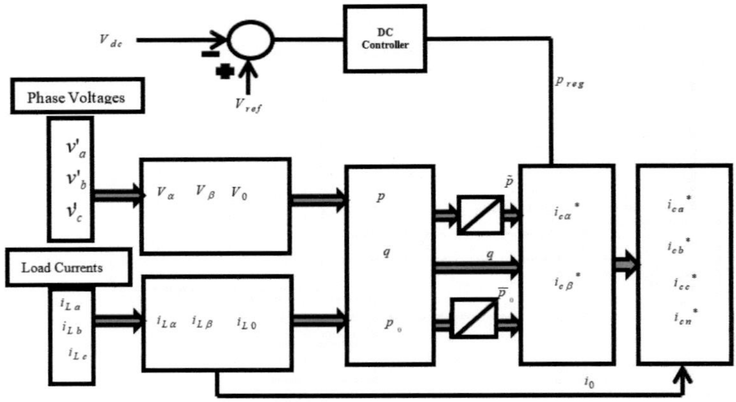

Figure 3.17 Modified p-q control theory.

$i'_c(\omega t) = \sin(\omega t + 2\pi/3)$ of Figure 3.16 are built up by the PLL circuit just calculating the time integral of the output ω of the PI controller. The PLL can reach a stable point of operation only if the input $p'_{3\varphi}$ of the PI controller has, in the steady state, a zero-average value, that is, $\overline{p'_{3\varphi}} = 0$. Moreover, the control circuit should minimize oscillations in $p'_{3\varphi}$ at low frequencies. The oscillating portion of $p'_{3\phi}$, where $p'_{3\phi} = p'_{3\phi} + p'_{3\phi}$ at low frequencies is not well attenuated by the PI controller and may bring instability to the PLL control circuit. The average three-phase power $P'_{3\phi} = p'_{3\phi}$ in terms of phasors, is given by:

$$P'_{3\varphi} = p'_{3\varphi} = 3V_{+1}I_{+1}\cos\varphi \,. \tag{3.57}$$

The above constraints are found only if ω equals the system frequency, and the current $i'_a(\omega t)$ becomes orthogonal to the fundamental positive-sequence component of the measured three-phase voltages v_a, v_b, *and* v_c. The PLL has only one stable point of operation, that is, $i'_a(\omega t)$ should lead 90° the phase voltage v_a. Since $i'_a(\omega t) = \sin(\omega t)$ in Figure 3.16 leads by 90° the fundamental positive-sequence component V_{+1} of the measured system voltages. Thus, the generated auxiliary current $i'_\alpha(\omega t) = \sin(\omega t - 2\pi/3)$ is in phase with V_{+1}. Figure 3.17 displays the block diagram for the modified p-q control theory.

3.10 Synchronous Reference Frame Theory

The SRF has a PI controller to control the voltage of the DC capacitor and a PLL circuit for vector positioning. Figure 3.18 displays the SRF technique's

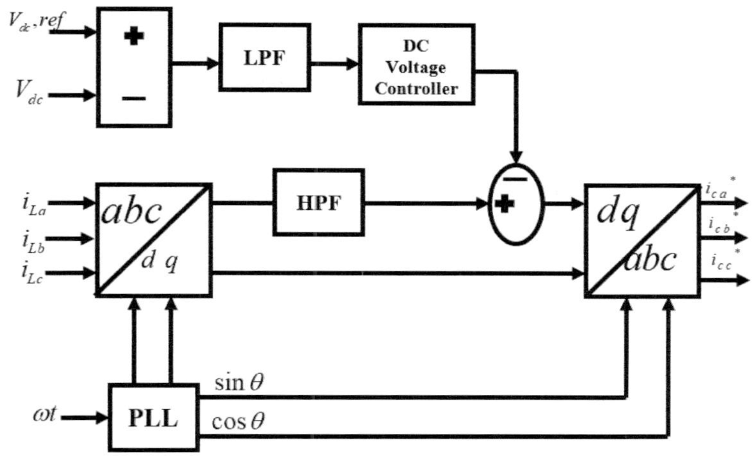

Figure 3.18 Block diagram of conventional SRF method.

traditional block diagram. The currents at the load side i_{La}, i_{Lb}, i_{Lc} in abc are transformed into the direct axis (*d*) and quadratic axis (*q*) which is two-phase rotating coordinates currents, i.e., $i_d - i_q$. This is given in eqn (3.58)

$$\begin{bmatrix} i_d \\ i_q \end{bmatrix} = \frac{2}{3}\begin{bmatrix} \sin\theta & \sin(\theta - 2\pi/3) & \sin(\theta + 2\pi/3) \\ \cos\theta & \cos(\theta - 2\pi/3) & \cos(\theta + 2\pi/3) \end{bmatrix}\begin{bmatrix} i_{La} \\ i_{Lb} \\ i_{Lc} \end{bmatrix}. \quad (3.58)$$

Power frequency currents are always synchronously rotated by the reference frame. Consequently, following conversion, time variant currents with power frequencies remain constant. Therefore, currents start separating instantaneously as DC and AC components. Reactive power correction and harmonic mitigation are two applications for the *d*-components. The output of the *d-q* transformation is largely dependent on the load current and the phase-locked loop's performance [55].

The vectorized power frequency ($\sin\theta$ and $\cos\theta$) for synchronization is supplied by the PLL circuit. A low-pass filter filters the Park-transformed currents, removing higher-order harmonic components while permitting power frequency components. By eliminating the DC steady state error, the PI controller keeps the VSI's capacitance voltage steady. The PI controller controls the DC voltage's settling time and dynamic response. To reduce the inverter losses by keeping the capacitor voltage constant, the real component of the *dq* current's positive-sequence power frequency is added to the loss current. The desired reference current (*a-b-c* stationary frame) is derived from $i_d - i_q$

rotating frame using inverse Park transformation as given in eqn (3.59).

$$\begin{bmatrix} i_{sa}{}^{*} \\ i_{sb}{}^{*} \\ i_{sc*} \end{bmatrix} = \begin{bmatrix} \sin\theta & \cos\theta \\ \sin(\theta - 2\pi/3) & \cos(\theta - 2\pi/3) \\ \sin(\theta + 2\pi/3) & \cos(\theta + 2\pi/3) \end{bmatrix} \begin{bmatrix} i_d \\ i_q \end{bmatrix}. \tag{3.59}$$

Through the use of an indirect PWM-current controller, the drawn reference current is connected to the actual compensation currents, resulting in the appropriate switching pulses for the inverter. However, a PLL circuit is necessary for the traditional SRF method. But the design of PLL circuit is complex, when multiple zero-crossing problems occur in the grid voltage. In contrast to the PLL circuit, the unit vector approach is straightforward and effective. The modified-SRF approach is combined with the unit vector.

Remarks: SRF theory uses Park transformation which is a rotational transformation that gives robust performance under dynamic loading conditions and transient conditions. PLL in conventional SRF theory makes the circuit complex. This can be overcome by unit vector template instead of PLL.

3.11 Modified Synchronous Reference Frame Theory

The structure depicted in Figure 3.19 is the modified-SRF. To provide the required reference current, it uses a rotational transformation, DC capacitor voltage regulation, and a simplified unit vector. Figure 3.20 shows the block diagram for the straightforward and effective unit vector approach. Synchronization is generated by sensing and computing the instantaneous grid voltages [56]. Using the Clarke transformation, three-phase grid voltages are instantly transformed into two-phase stationary $\alpha - \beta$ voltages, as shown in eqn (3.60).

$$\begin{aligned} v_{s\alpha} &= \frac{3}{2} v_m \sin(\omega t) \\ v_{s\beta} &= -\frac{3}{2} v_m \cos(\omega t). \end{aligned} \tag{3.60}$$

The power frequency's first-order digital low-pass filter has been employed to eliminate the $\alpha - \beta$ voltages. The percentage of *h*th-order grid voltage harmonics are condensed by a factor of $\sqrt{2/\left(h^2+1\right)}$. Higher-level harmonics, notches, and noise are eliminated as a result. The evaluated magnitude of the generated space vector is displayed in eqn (3.61).

$$\vec{V} = \vec{V}_{\alpha\beta} = v_{s\alpha} + j v_{s\beta} = \sqrt{(v_{s\alpha}{}^2 + v_{s\beta}{}^2)}. \tag{3.61}$$

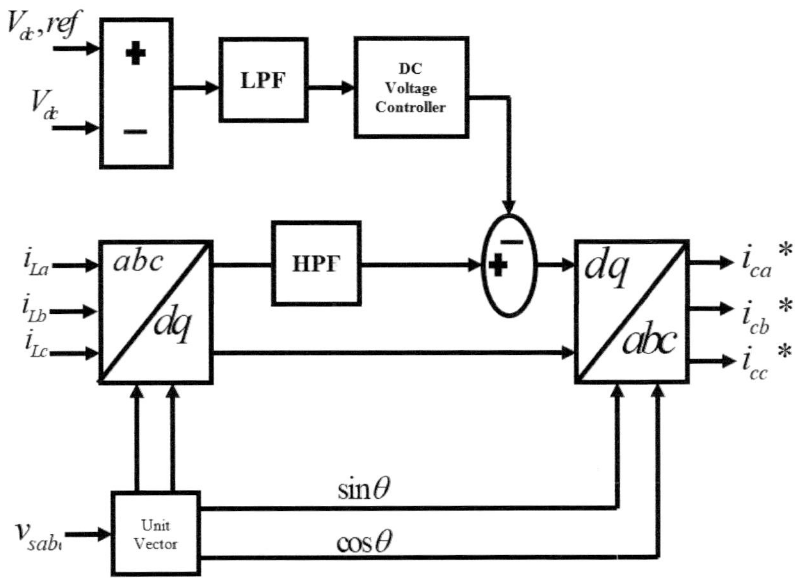

Figure 3.19 Block diagram of the modified-SRF method.

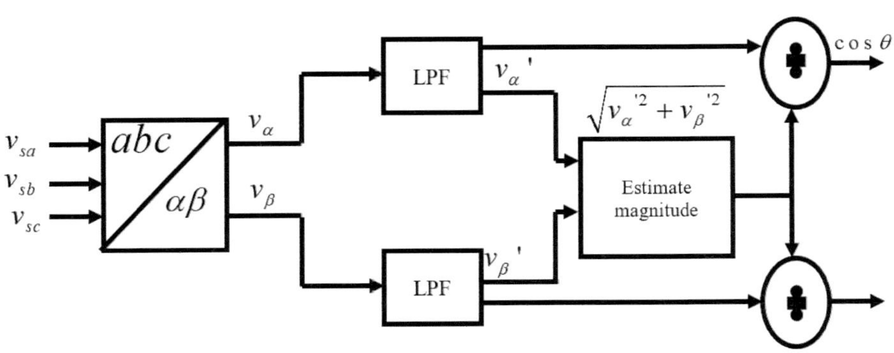

Figure 3.20 Block diagram of the unit vector generation.

As of the eqn (3.61), it has been demonstrated that the unit vectors are produced by dividing the $\alpha - \beta$ components by the space vector magnitude after converting the grid voltage to the $\alpha - \beta$ plane. Therefore, eqn (3.62) provides the unit vector generation.

$$\cos\theta = \frac{v_\alpha}{\sqrt{({v_{s\alpha}}^2 + {v_{s\beta}}^2)}}$$

$$\sin\theta = \frac{v_\beta}{\sqrt{\left(v_{s\alpha}{}^2 + v_{s\beta}{}^2\right)}}. \tag{3.62}$$

This unit vector method outcomes in mitigation of grid harmonics, noise, and notches.

Remarks: The PLL circuit is a necessary for traditional SRF method. However, the architecture of the PLL circuit is intricate. The drawback of conventional SRF is overcome by modified SRF. The updated SRF approach has made use of a straightforward and effective unit vector. The performance of the modified SRF technique is superior to that of the traditional SRF technique.

3.12 Proposed Averaging pq-SRF (Apq-SRF) Control Theory

The proposed Apq-SRF control's block diagram is displayed in Figure 3.21. The two compensating reference currents use the same function, but one is produced from the Park transformation and the other from the Clark transformation. The compensating reference currents are obtained from the p-q and SRF approaches independently.

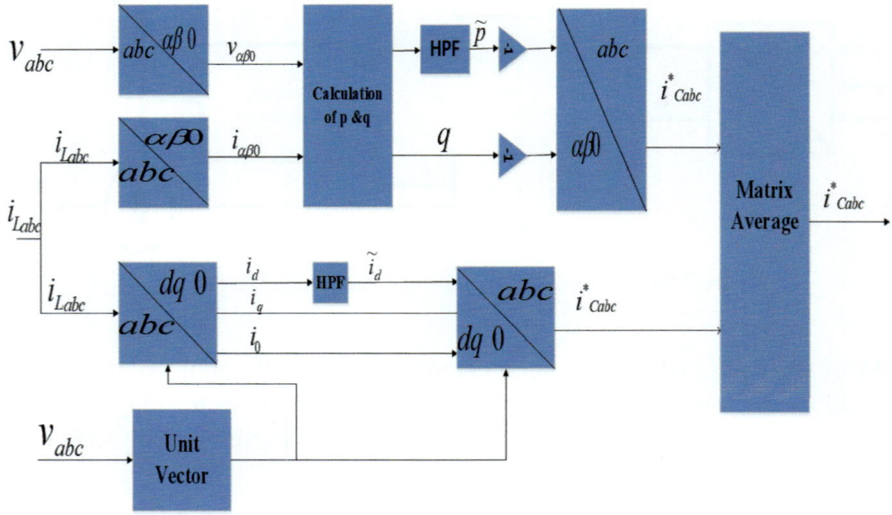

Figure 3.21 Block diagram of proposed Apq-SRF control.

Despite undergoing distinct changes, these two compensating reference currents have the perfect properties. Getting these two compensatory reference currents to correlate is a difficult challenge. There are other ways to determine the correlation between two distinct and similar equations, but the simplest and most straightforward approach is matrix averaging, which combines the characteristics of two distinct and similar equations. SRF and p-q are regarded as separate but related equations, and the two compensating reference currents are taken from them. Therefore, matrix averaging of p-q and SRF reference currents is carried out in this suggested study to obtain the characteristics of both p-q and SRF control methods with reduced complexity and improved efficiency in comparison to their separate applications.

Mathematical modeling of Apq-SRF control technique

The voltages and currents in stationary reference frame ($\alpha\beta 0$) is given by eqn (3.63) and (3.64).

$$\begin{bmatrix} v_0 \\ v_\alpha \\ v_\beta \end{bmatrix} = \sqrt{\frac{2}{3}} \begin{bmatrix} \frac{1}{\sqrt{2}} & \frac{1}{\sqrt{2}} & \frac{1}{\sqrt{2}} \\ 1 & -\frac{1}{2} & -\frac{1}{2} \\ 0 & \frac{\sqrt{3}}{2} & -\frac{\sqrt{3}}{2} \end{bmatrix} \begin{bmatrix} v_a \\ v_b \\ vc \end{bmatrix} \tag{3.63}$$

$$\begin{bmatrix} i_0 \\ i_\alpha \\ i_\beta \end{bmatrix} = \sqrt{\frac{2}{3}} \begin{bmatrix} \frac{1}{\sqrt{2}} & \frac{1}{\sqrt{2}} & \frac{1}{\sqrt{2}} \\ 1 & -\frac{1}{2} & -\frac{1}{2} \\ 0 & \frac{\sqrt{3}}{2} & -\frac{\sqrt{3}}{2} \end{bmatrix} \begin{bmatrix} i_{La} \\ i_{Lb} \\ i_{Lc} \end{bmatrix}. \tag{3.64}$$

The instantaneous real power, reactive power, and zero-sequence real power are computed using eqn (3.65) from the stationary reference voltages and currents.

$$\begin{bmatrix} p_0 \\ p \\ q \end{bmatrix} = \begin{bmatrix} v_0 & 0 & 0 \\ 0 & v_\alpha & v_\beta \\ 0 & v_\beta & -v_\alpha \end{bmatrix} \begin{bmatrix} i_0 \\ i_\alpha \\ i_\beta \end{bmatrix}. \tag{3.65}$$

The instantaneous real power (p) have two terms – fundamental ($\hat{p}$) and oscillating ($\tilde{p}$) real powers. The useful power is the fundamental real power ($\hat{p}$) which is always useful to the loads whereas oscillating real power ($\tilde{p}$) always oscillates between source and load and it is not useful. Similarly, instantaneous reactive power (q) also have fundamental and oscillating power, but both are not useful to the loads. Hence, oscillating real power ($\tilde{p}$) and reactive power (q) must be compensated by SAF, so negative symbol is assigned as in eqn (3.66).

The zero-sequence component of the $\alpha\beta 0$ transformation stays the same as the ABC frame, which is an intriguing feature.

$$\begin{bmatrix} i_{C\alpha} \\ i_{C\beta} \end{bmatrix} = \frac{1}{\sqrt{{v^2}_\alpha + {v^2}_\beta}} \begin{bmatrix} v_\alpha & v_\beta \\ -v_\beta & v_\alpha \end{bmatrix} \begin{bmatrix} -\tilde{p} + \Delta\hat{p} \\ -q \end{bmatrix}. \tag{3.66}$$

Equation (3.67), in the abc frame, presents the compensating reference currents. Power factor adjustment, reactive power compensation, and harmonic elimination are all handled by these compensating reference currents. These reference currents are ineffective in carrying out their necessary tasks when the supply is distorted and/or imbalanced. This is the main issue with the p-q control theory.

$$\begin{bmatrix} {i^*}_{Ca} \\ {i^*}_{Cb} \\ {i^*}_{Cc} \end{bmatrix} = \sqrt{\frac{2}{3}} \begin{bmatrix} \frac{1}{\sqrt{2}} & 1 & 0 \\ \frac{1}{\sqrt{2}} & -\frac{1}{2} & \frac{\sqrt{3}}{2} \\ \frac{1}{\sqrt{2}} & -\frac{1}{2} & -\frac{\sqrt{3}}{2} \end{bmatrix} \begin{bmatrix} -i_0 \\ i_{C\alpha} \\ i_{C\beta} \end{bmatrix}. \tag{3.67}$$

SRF control theory employs a rotational frame ($dq0$), also known as the Park transformation, in which all of the phasors rotate at the same speed, as opposed to p-q control. Equation (3.68) illustrates how the currents in the ($dq0$) frame are derived from the ABC frame.

At $t = 0$, the d-axis is aligned with the a-axis if the rotating frame is aligned with an axis at that time. The cosine-based Park transformation is the name given to this kind of Park transformation. In contrast, the q-axis is aligned with the a-axis if the rotating frame is 90 degrees behind an axis, or when $t = 0$. The sine-based Park transformation is another name for this kind of Park transformation.

Figure 3.22 displays the phasor relationship between $abc - \alpha\beta 0 - dq0$. Equation (3.68) illustrates the relationship that results when the rotating frame is aligned with the a-axis.

$$\begin{bmatrix} i_d \\ i_q \\ i_0 \end{bmatrix} = \frac{2}{3} \begin{bmatrix} \cos\theta & \cos(\theta - 120) & \cos(\theta + 120) \\ -\sin\theta & -\sin(\theta - 120) & -\sin(\theta + 120) \\ \frac{1}{2} & \frac{1}{2} & \frac{1}{2} \end{bmatrix}. \tag{3.68}$$

Direct current i_d have two components – fundamental component ($\hat{i}_d$) and oscillating component ($\tilde{i}_d$). The oscillating component ($\tilde{i}_d$) exchanges between source and load and must be extracted. Similarly, the quadrature

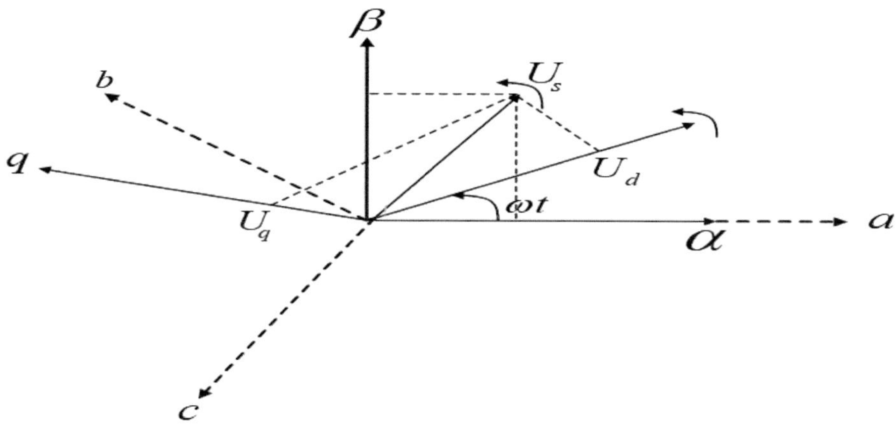

Figure 3.22 Phasor relationship between $abc - \alpha\beta - dq0$ frames.

current (i_q) also have two components. However, both of the components need to be removed because they are oscillatory in nature. Equation (3.69) provides the SRF compensating reference currents for SAF.

The PLL block is utilized to determine the rotational angle (θ) between d and α, but its circuit design is intricate.

$$\begin{bmatrix} i^*{}_{Ca} \\ i^*{}_{Cb} \\ i^*{}_{Cc} \end{bmatrix} = \sqrt{\frac{2}{3}} \begin{bmatrix} \cos\theta & -\sin\theta & 1 \\ \cos(\theta - 120) & -\sin(\theta - 120) & 1 \\ \cos(\theta + 120) & -\sin(\theta + 120) & 1 \end{bmatrix} \begin{bmatrix} \tilde{i}_d - i_{\text{loss}} \\ i_q \\ i_0 \end{bmatrix}. \tag{3.69}$$

It is evident that SRF control's zero-sequence current is identical to that of p-q control. Therefore, in the ABC reference frame, the zero-sequence current is likewise the same. It is among the intriguing aspects of changes. SRF control has the benefit of functioning effectively even in the presence of distorted and/or imbalanced supply conditions. When it comes to harmonic elimination, it works better than p-q control. The disadvantage of SRF control is that it is ineffective for load balancing and lacks a hint for reactive power compensation in the compensating reference current, as indicated by eqn (3.69).

Equations (3.67) and (3.69), which undergo two distinct transformations, share similar features. Matrix averaging is used to correlate the two equations in order to improve performance using the unique control approach known as Apq-SRF control. Equation (3.70) displays the combined reference current equation.

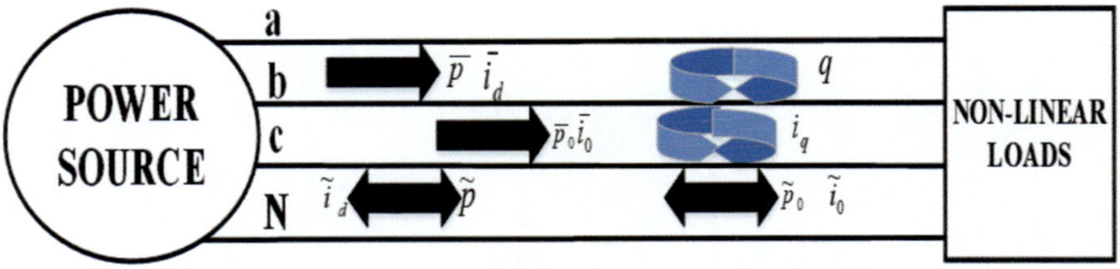

Figure 3.23 Power flow diagram of proposed Apq-SRF control.

$$\begin{bmatrix} i^*{}_{Ca} \\ i^*{}_{Cb} \\ i^*{}_{Cc} \end{bmatrix} = \sqrt{\frac{1}{6}} \begin{bmatrix} i_{c\alpha} + i_{c\beta} + \left(\tilde{i}_d - i_{\text{loss}}\right)\cos\theta - i_q \sin\theta + 0.3 i_0 \\ -0.5 i_{c\alpha} + \frac{\sqrt{3}}{2} i_{c\beta} + (\tilde{i}_d - i_{\text{loss}})\cos(\theta - 120°) - i_q \sin(\theta - 120) + 0.3 i_0 \\ -0.5 i_{c\alpha} - \frac{\sqrt{3}}{2} i_{c\beta} + (\tilde{i}_d - i_{\text{loss}})\cos(\theta + 120°) - i_q \sin(\theta + 120) + 0.3 i_0 \end{bmatrix}. \tag{3.70}$$

The power flow diagram of the proposed Apq-SRF control strategy is as shown in Figure 3.23. It can be observed that $\hat{p}$ and $\hat{i}_d$ are direct components; both can be combined to develop the new reference currents for SAF. $\hat{p}_0$ and $\hat{i}_0$ are also direct components like $\hat{p}$ and $\hat{i}_d$ but results in power loss via neutral conductor and hence, these must be eliminated along with oscillating components i_q,q,$\tilde{p}$, and $\tilde{i}_d$.

3.13 Current Control Techniques

The comparison of reference currents and sensing actual currents yields the three-phase error signals in PWM-VSI current control systems.

The IGBT/MOSFET employed in the SAF will produce the required switching pulses as a result.

The source current I_s or SAF current I_c can be used to construct one of two PWM-VSI control strategies.

- Controllers for direct current and indirect current

Supplying a nonlinear demand with power frequency real power and compensating for distribution network losses are the grid's primary responsibilities. Since the SAF is supposed to produce the load's anti-harmonics locally, the grid currents ought to be almost sinusoidal and in sync with the grid voltages. It is essential to supply additional actual power to deal with the SAF losses and to regulate and sustain the VSI capacitor voltage. Supplying a nonlinear demand with power frequency real power and compensating for distribution network losses are the grid's primary responsibilities. Since the SAF is supposed to produce the load's anti-harmonics locally, the grid currents ought to be almost sinusoidal and in sync with the grid voltages. It is essential to supply additional actual power to deal with the SAF losses and to regulate and sustain the VSI capacitor voltage.

3.13.1 Direct current control technique

To develop the direct current control scheme, seven analog signals (v_{sa} , v_{sb}, v_{sc} , V_{dc} , i_{la} , i_{lb}, i_{lc}) and three compensation currents (i_{ca}, i_{cb}, i_{cc}) are sensed and computed. The block diagram of the direct current control technique is as shown in Figure 3.24 (a). The PWM-VSI switching pulses are generated by relating the sensed three-phase SAF currents (i_{ca}, i_{cb}, i_{cc}) with reference counterpart's (i_{ca*}, i_{cb*}, i_{cc*}) currents [69].

3.13.2 Indirect current control technique

Four analog signals (v_{sa}, v_{sb}, v_{sc}, V_{dc}) and three actual source currents (i_{sa}, i_{sb}, i_{sc}) are sensed and computed to develop the indirect current control technique. The topology of the indirect current control scheme is as shown in Figure 3.24 (b). The PWM-VSI switching pulses are generated by relating

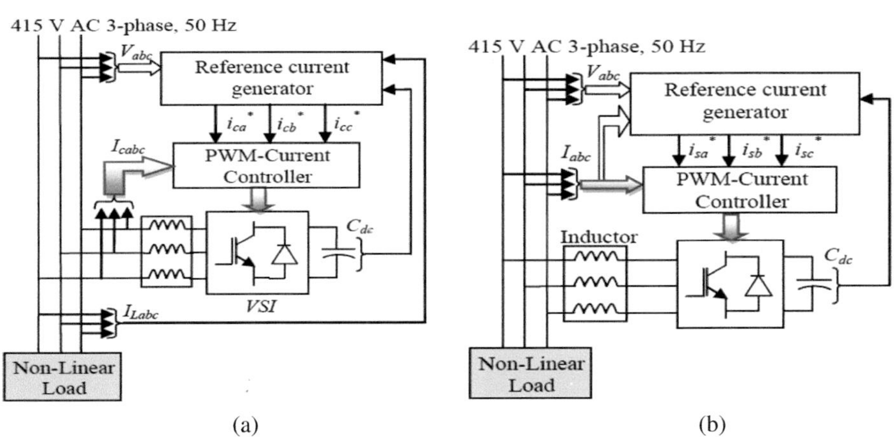

Figure 3.24 Block diagram of pulse generation (a) direct current controller and (b) indirect current controller.

Table 3.3 Comparison between indirect current control and direct current control.

Characteristics	Indirect current control	Direct current control
Reference signal	Voltage	Current
Knowledge of system parameters	Required	Not required
Influence by the variations of system parameters	More	Less
Response time	Slower	Faster
Control scheme	Feedback control/feedback control and feedforward control	Feedback control

the sensed grid currents (i_{sa}, i_{sb}, i_{sc}) with their reference currents (i_{sa*}, i_{sb*}, i_{sc*}).

The gate driver circuit will be used to process the PWM-VSI switching pulses in order to accomplish optical isolation and amplification [69]. PWM-VSI switching pulses for SAF are provided by the gate driver circuit. The SAF should stabilize the dc capacitor voltage and reduce load harmonics based on the PWM-VSI switching pulses. Thus, only clean currents from the ac mains can be drawn by the SAF system with polluting loads.

Table 3.3 provides an overview of the distinctions between direct current control and indirect current control. Compared to the indirect current control approach, direct current control offers a faster dynamic response and more resilience. As a result, the SAF in this study uses the direct current control technique.

3.14 Modeling of PWM-VSI

The PWM-VSI input is connected to three-phase grids through interface impedance; output is connected to dc-link capacitor. The voltages v_{ca}, v_{cb}, v_{cc} can be formulated with dc capacitor voltage V_{dc} and switching functions states, i.e., ON/OFF status of the switches of each leg S_A, S_B, *and* S_C as

$$
\begin{aligned}
v_{ca} &= \frac{V_{dc}}{3}\left(2S_A - S_B - S_C\right) \\
v_{cb} &= \frac{V_{dc}}{3}\left(-S_A + 2S_B - S_C\right) \\
v_{cc} &= \frac{V_{dc}}{3}\left(-S_A - S_B + 2S_C\right).
\end{aligned} \tag{3.71}
$$

The compensating currents i_{ca}, i_{cb}, *and* i_{cc} going through the filter elements (R_c, L_c) are given by eqn (3.72)

$$\begin{aligned} i_{ca} &= -\left(\frac{R_c}{L_c}\right) i_{ca} + (v_{sa} - v_{ca})/L_c \\ i_{cb} &= -\left(\frac{R_c}{L_c}\right) i_{cb} + (v_{sb} - v_{cb})/L_c \\ i_{cc} &= -\left(\frac{R_c}{L_c}\right) i_{cc} + (v_{sc} - v_{cc})/L_c. \end{aligned} \tag{3.72}$$

The dc-link current is expressed with compensating currents i_{ca}, i_{cb}, *and* i_{cc} and the status of switching, i.e., S_A, S_B, *and* S_C

$$I_{dc} = i_{ca}S_A + i_{cb}S_B + i_{cc}S_C. \tag{3.73}$$

From this, dc capacitor voltage can be expressed as

$$V_{dc} = \frac{1}{C_{dc}}\left(i_{ca}S_A + i_{cb}S_B + i_{cc}S_C\right). \tag{3.74}$$

To provide the inverter with actual power, the dc capacitor serves as an energy storage component. Under dynamic load-changing conditions, it provides the difference between the real power of the load and the supply. When everything is operating normally, the source real power should equal the load real power + SAF losses. The grid current will be in phase with the grid voltage if the SAF compensates for both the reactive and harmonic current.

3.15 Triangular Carrier Current Controller

For the SAF PWM-VSI inverter, it is one of the well-known techniques for creating gate pulses. It is accomplished independently for every stage. It produces pulses that are sent straight to the three-phase VSI system [57].

For A-phase, the actual compensation current is represented as i_{ca} and reference compensation current is represented as i_{ca*}. In similar manner, for B-phase and C-phase currents are expressed as i_{cb} *and* i_{cb*} and i_{cc} *and* i_{cc*}. Figure 3.25 represents the topology of the A-phase followed triangular-carrier current controller.

The triangle-shaped carrier signal and the error signal are connected to produce the necessary switching pulses. The bottom inverter switch is ON

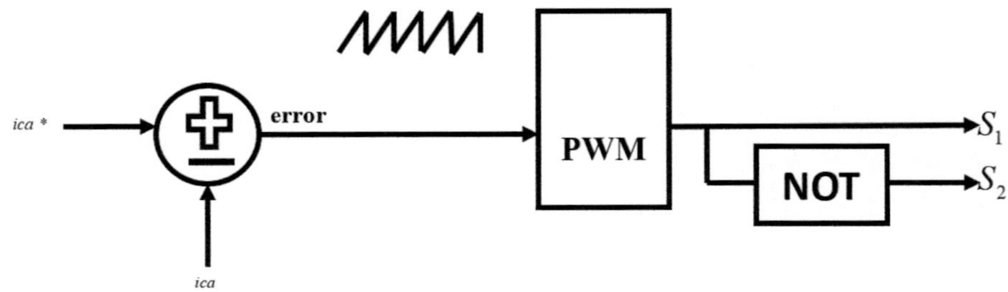

Figure 3.25 Block diagram of triangular carrier current controller.

if the error signal is less than the carrier signal, and the upper inverter switch is ON when the error signal is greater than the triangular carrier signal. The inverter switch's switching frequency and the carrier signal's frequency are same [57]. It offers less current ripple and superior harmonic mitigation as compared to alternative switching techniques. Lower bandwidth and proportional gain overshoot problems are this controller's disadvantages.

3.16 Fixed hysteresis current controller

Upon comparing the error $e(t)$ to the fixed-hysteresis bands, switching pulses are generated. Figure 3.26 (a) displays the topology of this approach. The discrepancy between the reference and actual compensation currents is called the error. As the error exceeds the upper limit of the fixed hysteresis band, the inverter arm's superior switch is turned OFF and the lower switch is switched ON, causing the current to decay. Figure 3.26 (b) illustrates the ON and OFF switching pulses to VSI. Similarly, the superior switch of VSI is ON and the

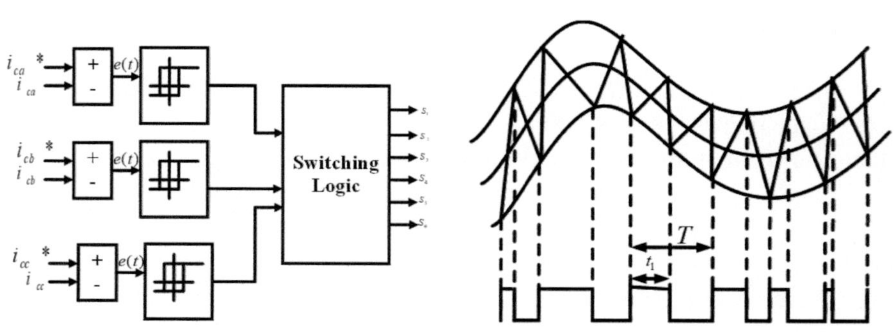

Figure 3.26 Block diagram of (a) two-level HCC and (b) two-level switching pattern.

lower switch is OFF if the erroneous signal crosses the lower limit of the specified hysteresis band. The current will therefore return to the hysteresis band, forcing the compensating current to follow the reference compensation current inside the fixed hysteresis band.

Due to the simplicity of the circuit, SAF applications are the primary use for these controllers. To generate the required switching pulses, it exclusively requires positive and negative dc voltages–zero dc voltages are not used.

One disadvantage of a set hysteresis band is that it generates sideband harmonics at the switching frequency. Performance is better with an adaptive hysteresis current controller than with a fixed hysteresis band [58].

3.17 Adaptive-hysteresis Current Controller

The adaptive-HCC-based switching pulse generator block design for a PWM-voltage source inverter is displayed in Figure 3.27 (a). Based on indirect current control theory, the adaptive-hysteresis band controller that Bose suggested for the machine drive system is used in this instance for three-phase, three-wire SAF [59].

The lower switch is activated when the current error $e(t)$ surpasses the adaptive hysteresis band's upper limit. If the error current crosses the upper switch is activated at the adaptive-hysteresis band's lower limit. As a result, the reference current must be followed by the actual current inside the hysteresis band. The ac-side power flow of the SAF conditioner is represented by single-line switching in Figure 3.27 (b). For the switching intervals t_1 and

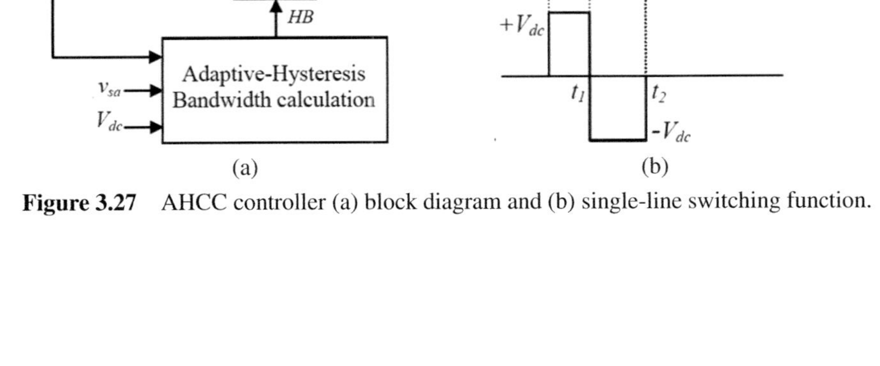

Figure 3.27 AHCC controller (a) block diagram and (b) single-line switching function.

t_2, the following formulas can be expressed.

$$\begin{aligned} \frac{di_{sa}{}^{+}}{dt} &= \frac{1}{L}\left(V_{dc} - v_s\right) \\ \frac{di_{sa}-}{dt} &= \frac{1}{L}\left(V_{dc} + v_s\right), \end{aligned} \tag{3.75}$$

where L is phase inductance, $(i_{sa}{}^{+})$ and $(i_{sa}{}^{-})$ are the respective rising and falling current segments.

The following equations can be obtained in the hysteresis-band with the corresponding switching intervals from Figure 3.27 (b).

$$\frac{di_{sa}+}{dt}t_1 - \frac{di_{sa}{}^{*}}{dt}t_1 = 2HB \tag{3.76}$$

$$\frac{di_{sa}-}{dt}t_2 - \frac{di_{sa}{}^{*}}{dt}t_2 = -2HB \tag{3.77}$$

$$t_1 + t_2 = T_c = 1/f_c, \tag{3.78}$$

where t_1 and t_2 are the respective switching intervals, and f_c is the modulation frequency. Combining the above two equations

$$\frac{di_{sa}+}{dt}t_1 + \frac{di_{sa}-}{dt}t_2 - \frac{di_{sa}{}^{*}}{dt}\left(t_1 + t_2\right) = 0. \tag{3.79}$$

3.18 Adaptive-fuzzy-hysteresis Current Controller

To enhance SAF performance without precise knowledge of the SAF parameters, a hysteresis band with fuzzy controller is employed. Adaptive-fuzzy-HCC is represented as follows in Figure 3.28 (a). The adaptive-fuzzy-hysteresis bandwidth is modulated by grid voltage and slope reference compensation current. The slope of the reference compensation current $dic_a * /dt$ and grid voltage v_s are used as inputs for fuzzy processing as shown in Figure 3.28 (b). The output of fuzzy controller is the adaptive-fuzzy-hysteresis band *HB* [60]. Figure 3.29 displays the hysteresis bandwidth computation for the adaptive-fuzzy-hysteresis current controller.

The fuzzy logic is having five linguistic variables (NB, NM, ZE, PM, and PB) for each input variable i.e., slope of the reference current $di_{sa}{}^{*}/dt$ and supply voltage v_s and output variable is adaptive-fuzzy hysteresis band *HB*. Positive very small (PVS), positive small (PS), positive medium (PM), positive large (PB), and positive very big (PVB) are the output variables, as seen in Figure 3.30.

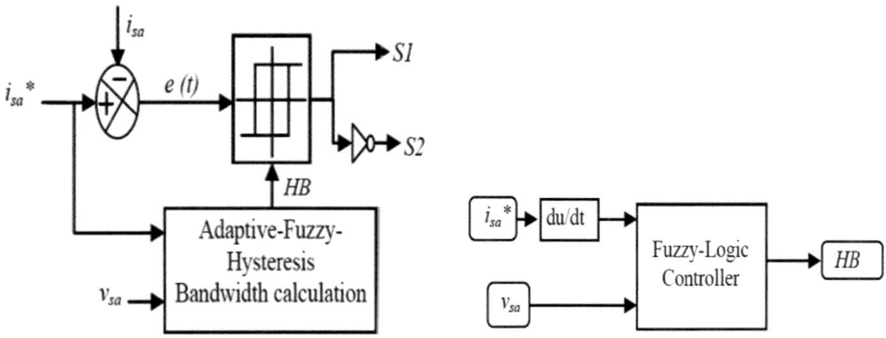

(a) (b)

Figure 3.28 AFHCC controller (a) block diagram and (b) fuzzy processing.

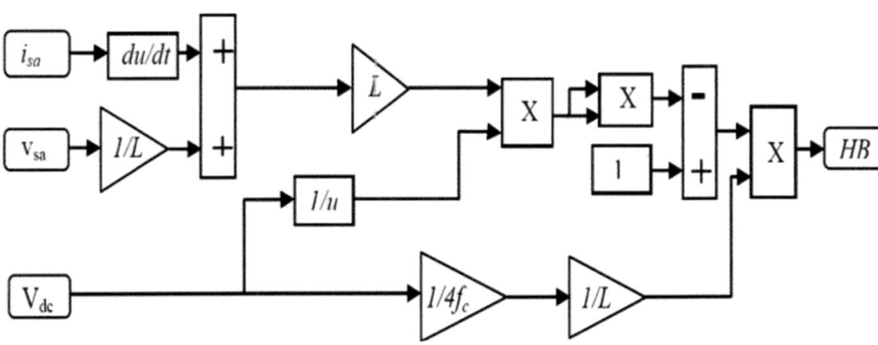

Figure 3.29 Hysteresis bandwidth calculation.

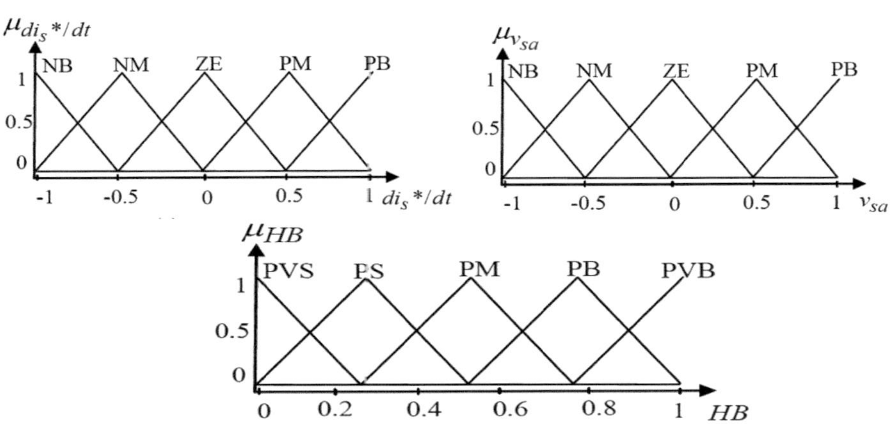

Figure 3.30 Membership functions for the input and variables di_{sa}^*/dt, v_{sa}, and HB.

Table 3.4 Fuzzy logic rules

$di_{sa}{}^*/dt$ \ v_{sa}	NB	NM	ZE	PM	PB
NB	PB	PM	PM	PM	PB
NM	PB	PM	PS	PM	PB
ZE	PVB	PM	PVS	PM	PVB
PM	PB	PM	PS	PM	PB
PB	PB	PM	PM	PM	PB

- The method of triangular membership functions is applied.
- A min-operator of the Mamdani type has been utilized for implication.
- For defuzzification, the centroid approach has been applied.

It is necessary for the rule evaluator to translate the fuzzy logic rules into language variables. With regard to adaptive-fuzzy-hysteresis current controllers, Table 3.4 lists the 25 rules that are utilized.

The HB is expressed as HB_a, HB_b, and HB_c for three-phase symmetrical operation, with a 120 degree phase difference and the same value. Further, fuzzy logic-based adaptive hysteresis bandwidth *HB* has to retain the modulation frequency f_c quasi constant. This controller's benefit is that it enhances SAF and lowers VSI switching losses [60].

3.19 Results and Discussion

The SAF is implemented with existing and proposed reference current extraction techniques, various DC-link voltage controllers and pulse-generating

Table 3.5 SAF performance with unit vector control.

Load	Harmonic controller	DC-link controller	Condition	Type of pulse generator				
				PWM	HCC	AHCC	AFHCC (type-1)	AF-HCC (type-2)
Diode bridge rectifier with RL load	Unit vector	PI	Steady State	4.96	4.2	4.1	4	3.9
			Transient	4.9	4.12	3.99	3.91	3.8
		Type-1 fuzzy	Steady State	4.8	4.01	3.96	3.83	3.77
			Transient	4.74	3.98	3.92	3.8	3.71
		Type-2 fuzzy	Steady state	4.7	3.96	3.9	3.75	3.67
			Transient	4.61	3.9	3.85	3.72	3.6

schemes. The performance of the SAF is evaluated separately for each reference current extraction technique in conjunction with different DC-link voltage controllers and pulse-generating schemes. In all the cases, the THD levels are within 5% and fall well below the IEE-519-2022 standards.

Table 3.5 shows the SAF performance with unit vector associated with different DC-link voltage controllers and pulse-generating schemes.

Table 3.6 shows the SAF performance with Fryze control associated with different DC-link voltage controllers and pulse-generating schemes.

Table 3.7 shows the SAF performance with p-q control associated with different DC-link voltage controllers and pulse-generating schemes.

Table 3.6 SAF performance with Fryze control.

Load	Harmonic controller	DC-link controller	Condition	Type of pulse generator				
				PWM	HCC	AHCC	AFHCC (type-1)	AF-HCC (type-2)
Diode bridge rectifier with RL load	Fryze	PI	Steady State	4.82	4.75	4.65	4.55	4.5
			Transient	4.8	4.7	4.61	4.59	4.52
		Type-1 fuzzy	Steady State	4.75	4.67	4.6	4.52	4.45
			Transient	4.71	4.65	4.59	4.49	4.4
		Type-2 fuzzy	Steady state	4.69	4.53	4.45	4.41	4.38
			Transient	4.6	4.51	4.44	4.38	4.31

Table 3.7 SAF performance with p-q control.

Load	Harmonic controller	DC-link controller	Condition	Type of pulse generator				
				PWM	HCC	AHCC	AFHCC (type-1)	AF-HCC (type-2)
Diode bridge rectifier with RL load	*p-q*	PI	Steady State	3.99	2.94	1.99	2.22	2.22
			Transient	3.24	1.64	1.27	1.20	1.20
		Type-1 fuzzy	Steady State	3.85	1.28	1.25	1.23	1.22
			Transient	4.35	1.80	1.89	1.78	1.76
		Type-2 fuzzy	Steady state	2.20	1.57	1.64	1.63	1.62
			Transient	1.20	1.17	1.19	1.15	1.12

Table 3.8 shows the SAF performance with modified p-q control associated with different DC-link voltage controllers and pulse-generating schemes.

Table 3.9 shows the SAF performance with SRF control associated with different DC-link voltage controllers and pulse-generating schemes.

Table 3.10 shows the SAF performance with modified SRF control associated with different DC-link voltage controllers and pulse-generating schemes.

Table 3.8 SAF performance with modified p-q control.

Load	Harmonic controller	DC-link controller	Condition	Type of pulse generator				
				PWM	HCC	AHCC	AFHCC (type-1)	AF-HCC (type-2)
Diode bridge rectifier with RL load	Modified *p-q*	PI	Steady State	3.45	2.08	2.05	2.08	2.05
			Transient	2.85	1.18	1.20	1.36	1.35
		Type-1 fuzzy	Steady State	3.87	1.12	1.9	1.96	1.35
			Transient	4.26	1.1	1.85	1.82	1.28
		Type-2 fuzzy	Steady state	3.6	1.08	1.82	1.6	1.21
			Transient	3.55	1.1	1.8	1.56	1.28

Table 3.9 SAF performance with SRF control.

Load	Harmonic controller	DC-link controller	Condition	Type of pulse generator				
				PWM	HCC	AHCC	AFHCC (type-1)	AF-HCC (type-2)
Diode bridge rectifier with RL load	SRF	PI	Steady State	3.45	3.3	3.15	3.13	3.12
			Transient	3.4	3.21	3.13	3	2.98
		Type-1 fuzzy	Steady State	3.35	3.15	3.12	2.95	2.92
			Transient	3.32	3.12	3.1	2.9	2.9
		Type-2 fuzzy	Steady state	3.27	3.15	3.05	2.85	2.8
			Transient	3.12	3.1	2.97	2.8	2.85

Table 3.10 SAF performance with modified SRF control.

Load	Harmonic controller	DC-link controller	Condition	Type of pulse generator: PWM	HCC	AHCC	AFHCC (type-1)	AF-HCC (type-2)
Diode bridge rectifier with RL load	Modified SRF	PI	Steady State	3.35	3.2	3.1	3	2.98
			Transient	3.3	3.15	3.05	2.96	2.85
		Type-1 fuzzy	Steady State	3.27	3.1	3	2.9	2.8
			Transient	3.2	3.05	2.96	2.88	2.78
		Type-2 fuzzy	Steady state	3.16	2.98	2.9	2.82	2.6
			Transient	3.1	2.92	2.9	2.72	2.62

Table 3.11 SAF performance with proposed Apq-SRF control.

Load	Harmonic controller	DC-link controller	Condition	Type of pulse generator: PWM	HCC	AHCC	AFHCC (type-1)	AF-HCC (type-2)
Diode bridge rectifier with RL load	Proposed Apq-SRF	PI	Steady State	3.27	3.18	3.09	2.91	2.85
			Transient	3.23	3.1	2.98	2.89	2.82
		Type-1 fuzzy	Steady State	3.2	3.05	2.9	2.82	2.78
			Transient	3.15	3	2.85	2.8	2.72
		Type-2 fuzzy	Steady state	3.1	2.9	2.8	2.75	**2.55**
			Transient	3.05	2.85	2.78	2.7	**2.52**

Table 3.11 shows the SAF performance with modified SRF control associated with different DC-link voltage controllers and pulse-generating schemes.

3.20 Contributions

✓ Introduced a novel reference current extraction control strategy i.e., Apq-SRF technique for the SAF.

✓ The proposed Apq-SRF technique is tested with existing DC-link voltage regulators and pulse-generating schemes. The performance is evaluated.

✓ The comparative analysis of SAF performance has been done among various reference current extraction techniques, DC-link voltage regulators, and pulse-generating schemes.

3.21 Summary

In this chapter, an energy-efficient power source can operate with a current source or a voltage source inverter. The PWM-voltage source inverter is well-suited for efficient SAF applications due to its simple design, low cost, and ability to be expanded into a multilevel inverter for higher power ratings. It is possible to construct a single-phase or three-phase SAF to accommodate various non-linear load types. Three-phase SAF is utilized in industrial settings, while single-phase SAF is frequently utilized in residential settings. There is discussion of the SAF idea as well as the traits of harmonic elements and compensating schemes. This chapter has examined the design of SAF elements.

This chapter elaborates and derives the time-domain methods for reference current extraction for SAF. The Apq-SRF methodology, a unique reference current extraction method, is created and put into practice. The fuzzy logic of PI/type-1 and type-2 By regulating the DC capacitor voltage of VSI, DC-link voltage control approaches have been employed to forecast the intended reference current. Under conditions of distorted supply, the traditional p-q hypothesis is ineffective. Even when grid voltages are distorted and out of balance, the modified p-q control method remains effective. However, compared to the traditional p-q control approach, the modified p-q control technique merits a longer working time.

The Fryze control technique does not require Clarke transformation, making this control strategy straightforward and reducing computation time. However, this method lacks a logic expression for calculating the instantaneous reactive power and compensating for non-linear loads' reactive power demands.

Accurate PLL circuit design is necessary for the SRF approach, but PLL circuit design is difficult. The improved SRF approach substitutes a basic unit vector for the PLL circuit.

In both steady and abrupt load-switching scenarios, as well as balanced, unbalanced, and distorted supply situations, the suggested Apq-SRF control provides the best and most effective performance when compared to traditional p-q and SRF systems.

Additionally, the PWM-VSI pulse production systems that generate the necessary gating pulses for VSI are presented in this chapter. This chapter introduces the TCCC, fixed-HCC, adaptive-HCC, and adaptive-fuzzy-HCC design methodologies.

The TCCC approach is relatively easy to use; it only requires a fixed frequency and amplitude. The error current, or the difference between the reference and real currents, is compared with the triangular-carrier signal. However, the constant frequency of the TCCC approach reduces the SAF operation and results in higher switching losses.

The fixed-HCC design is very simple in design and implementation is easy. A drawback of this technique is that it exhibits several undesirable characteristics, such as the switching frequency varying within a certain band limit.

Fixed HCC's shortcomings are addressed by adaptive-HCC. In accordance with the instantaneous current variations, the adaptive-HCC fluctuates with the bandwidth. Adaptive-HCC's disadvantage is that because of its high switching frequency, it has higher switching losses. The adaptive-fuzzy-HCC overcomes adaptive HCC's shortcomings.

The fuzzy logic controller is used in the adaptive-fuzzy-HCC technique to calculate the bandwidth without having precise information of the SAF parameters. When compared to previous approaches, the adaptive-fuzzy-HCC technique has the benefit of having lower switching losses and superior SAF performance.

4

Comparative Analysis

4.1 Introduction

This chapter compares the operation of the shunt hybrid active filter (SHAF) with various SAF reference current extraction techniques, DC-link voltage controllers and pulse-generating schemes and tuned passive filters. Figure 4.1 shows the SHAF configuration with different types of SAF control techniques and tuned passive filters. In this research, single-tuned and double-tuned passive filters are designed and investigated along with SAF reference current extraction techniques, DC-link voltage controllers and pulse generators. The SAF consists of a controller for maintaining the constant DC-link voltage, controller for reference current extraction and generating switching pulses. Chapter 2 investigates the various single-tuned and double-tuned passive filters; Chapter 3 investigates the various reference current extraction methods, DC-link voltage regulators, and direct method of switching control techniques. This chapter mainly focuses on comparative analysis of the SHAF control techniques, single- and double-tuned SHAF with the help of MATLAB Simulink. The VSI-based SAF is having six power switches with energy-stored capacitor and is connected to the PCC through interface inductor. The work of interface inductor is to suppress harmonic currents of higher order due to switching of power switches. The SHAF utilizes same simulation parameters for all control techniques of SAF and tuned passive filters. Each reference current extraction technique is linked to different switching control methods, DC-link voltage controllers, and tuned passive filters, which are analyzed separately. The SAF, single- and double-tuned SHAF are tested with two diode rectifier loads under steady and transient state conditions. The various controls for reference current extraction, DC-link voltage, and PWM generation along with tuned passive filters are shown in Figure 4.1.

The distorted load current waveform without any harmonic filter under steady and transient state is as shown in Figure 4.2. It is to observe that one

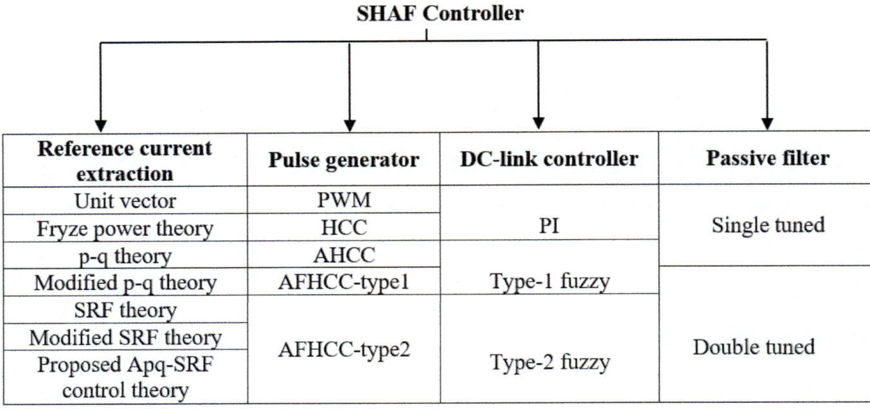

Figure 4.1 Subdivision of SHAF controller.

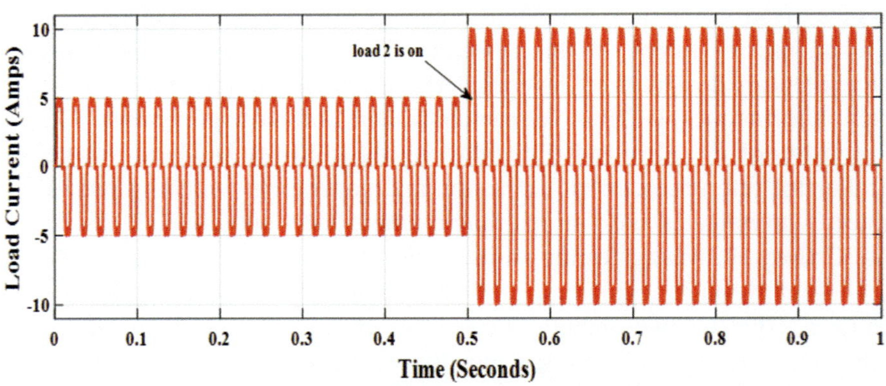

Figure 4.2 Distorted load currents without SHAF.

uncontrolled diode bridge rectifier acts as a fixed load connected at $t = 0$ sec. and other uncontrolled bridge rectifier is switched on at 0.3 sec. Thus, whole load acts as a dynamic load.

The THD of load current without any harmonic filter under steady state condition is as shown in Figure 4.3. It has been observed that, the THD is 26.12% which is beyond the limits of IEE-519-2022 standards.

The THD of load current without any harmonic filter under transient state condition is as shown in Figure 4.4. It has been observed that the THD is 26.01% which is also beyond the limits of IEE-519-2022 standards.

The grid voltage without any harmonic filter is as shown in Figure 4.5 and corresponding THDs under steady and transient state conditions are shown in

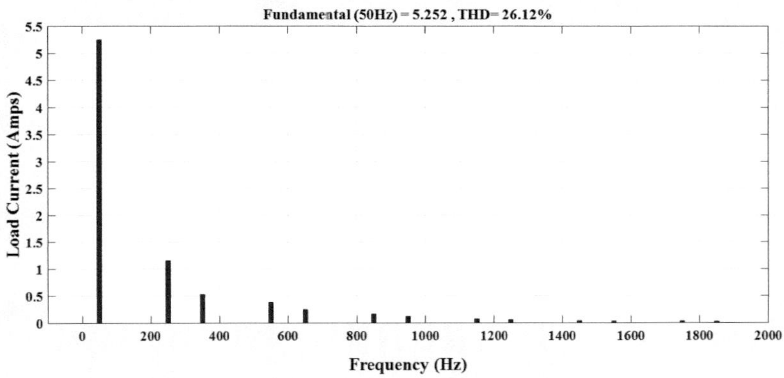

Figure 4.3 THD of load current without SHAF under steady state condition.

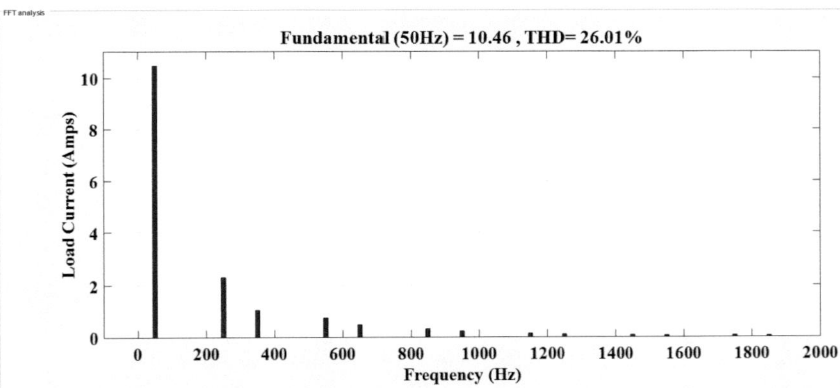

Figure 4.4 THD of load current without SHAF under steady state condition.

Figure 4.6 and Figure 4.7, respectively. It has been observed that there is no significant distortion in grid voltage and it falls within the IEE-519-2022 standards. The THD of grid voltage under steady state condition is 0.28 and transient state condition is 0.56.

The real power demand, reactive power demand, and the power factor of the system without any harmonic filter under steady and transient state conditions is as shown in Figure 4.8. It has been observed that, all these are oscillating in nature and oscillating powers leads to power loss. The role of harmonic filters is to minimize oscillations in real and reactive power, while also maintaining a unity power factor.

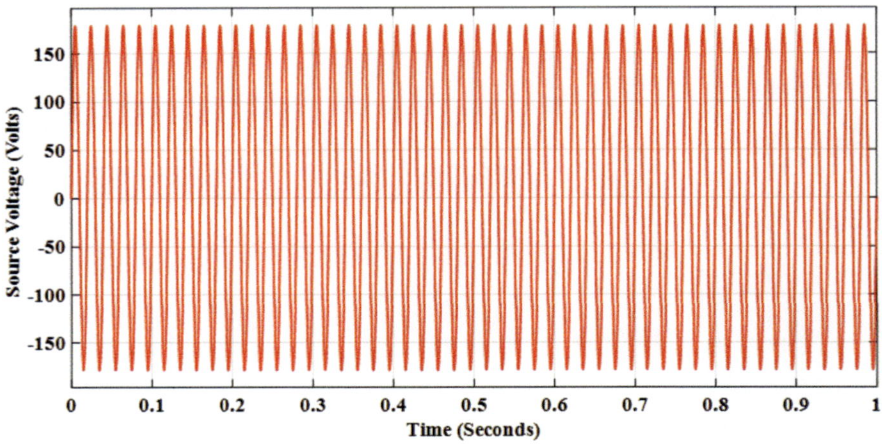

Figure 4.5 Grid voltage without SHAF.

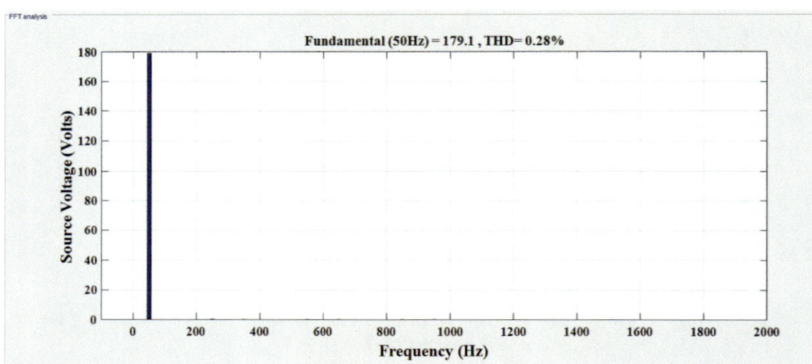

Figure 4.6 THD of grid voltage without SHAF under steady state condition.

4.2 Unit Vector Control Theory

The AC-main supply feeds two uncontrolled diode bridge rectifiers with R-L loads, first load is fixed and second load is on with a delay of 0.5 sec. Thus, it forms like a dynamic load. The grid current pulls a non-sinusoidal shape waveform due to the non-linear bridge rectifier loads. It means that, current is having fundamental and harmonic current components. To compensate the current harmonics, SAF is connected in parallel to the load at PCC that injects the anti-harmonics and make the supply current as sinusoidal. The SAF comprises six-IGBTs, DC-link capacitor, interface-filter, reference current extraction controller, and switching pulse generator. The

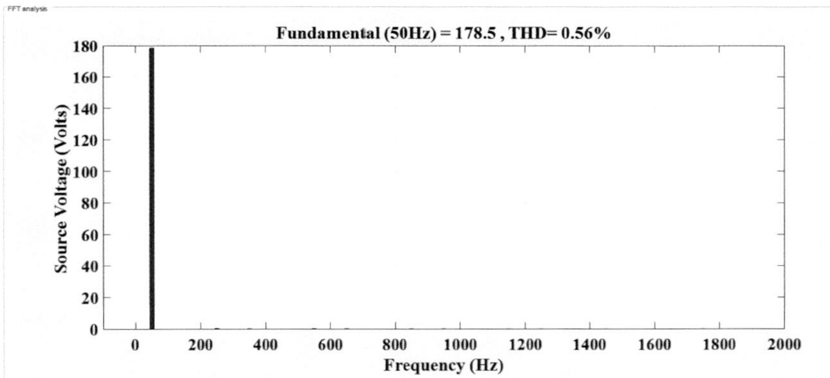

Figure 4.7 THD of grid voltage without SHAF under transient state condition.

reference current extraction is developed from unit current vector along with PI, type-1 fuzzy and type-2 fuzzy DC-link voltage controllers and TCCC, HCC, AHCC, AFHCC type-1 and AFHCC type-2 switching schemes is as shown in Figure 4.9. In order to anticipate the desired amplitude of maximum current by using DC VSI capacitor voltage, the PI, type-1 fuzzy and type-2 fuzzy DC-link voltage controllers have been used. The maximum amplitude get multiply with the unit vector current templates to produce the desired reference current. The produced reference current is exactly in track with actual SAF current to produce the switching pulses using PWM/HCC/AHCC/AFHCC type-1/AFHCC type-2 switching schemes.

The SHAF performance basically depends on the design and characteristics of PWM/HCC/AHCC/AFHCC type-1/AFHCC type-2 switching pulse generation techniques. The switching losses are more in PWM and HCC because of high frequency operation and also add high frequency ripples in produced current. This is overcome by means of AHCC techniques. With AHCC the bandwidth changes as per the instantaneous current changes. AFHCC enhances the PWM performance and also it gives an extra additional cushion for increase in switching frequency as per the system needs. The AFHCC greatly diminishes the switching losses and ripples in current. Additionally, the SHAF performance also depends on the effective design of single- or double-tuned passive filters. The single- and double-tuned passive filters will reduce the kVA size of the SAF drastically. Reduced SAF size gives the low cost and better performance along with tuned passive filters. From the literature, it has been observed that, the double-tuned SHAF gives the better performance as compared with the single-tuned SHAF.

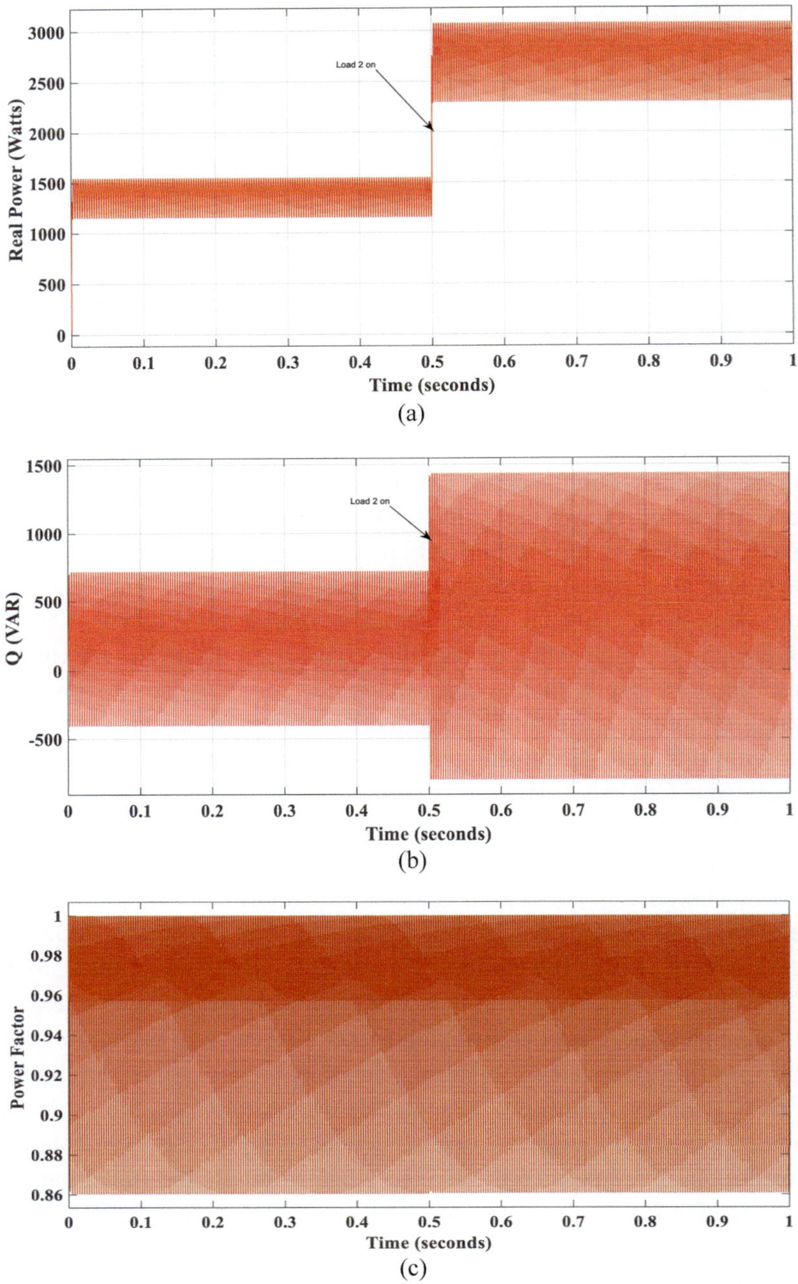

Figure 4.8 (a) Real power demand, (b) reactive power demand and (c) power factor without SHAF under steady and transient states.

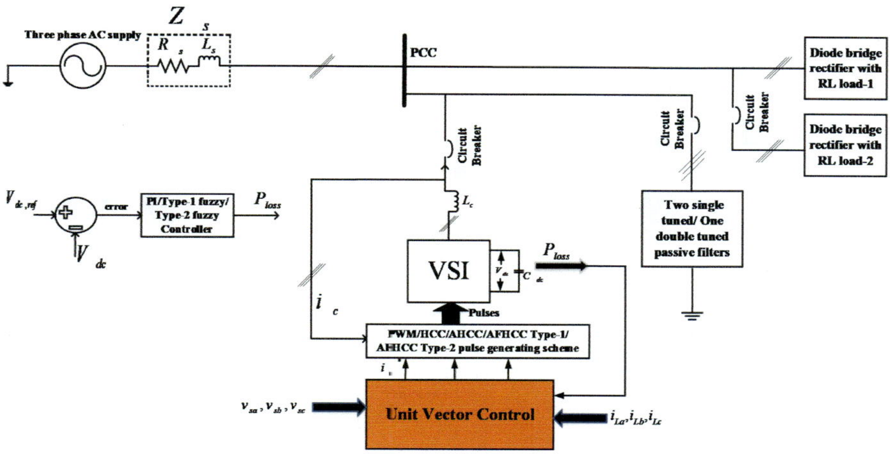

Figure 4.9 Unit vector-controlled shunt hybrid active filter.

4.2.1 Case-1 – PI DC-link voltage controller

PI DC-link voltage controller with different pulse-generating schemes and tuned passive filters is simulated for SHAF under two uncontrolled diode bridge rectifiers with RL-load conditions. The waveforms in Figure 4.10 are concentrated on PI DC-link voltage controller with PWM/HCC/AHCC/AFHCC type-1/AFHCC type-2 techniques respectively. Figure 4.10 (a) shows the load current without SHAF. It is observed that the load current is having power frequency as well as harmonics due to the non-linear loads (diode bridge rectifier loads). The SHAF when switched on at 0.1 sec provides the desired compensating current to cancel out the harmonic current at PCC as shown in Figure 4.10 (b). By injecting equal but opposite harmonic components using SAF and providing lower impedance path for lower-order current harmonics using either two single and one double-tuned passive filter, the harmonic compensation has been achieved. Figure 4.10 (c) shows the grid current after harmonic compensation. These waveforms are achieved under uncontrolled bridge rectifier load with steady-state and transient state conditions. The THD measurement of source/load current without filter under steady state condition is 26.12% and under transient condition is 26.01%.

The grid current of phase A is shown in Figure 4.11, when SAF with PI DC-link voltage controller and HCC switching scheme is connected to the double-tuned passive filter at 0.1 sec. It has been noticed that, the current

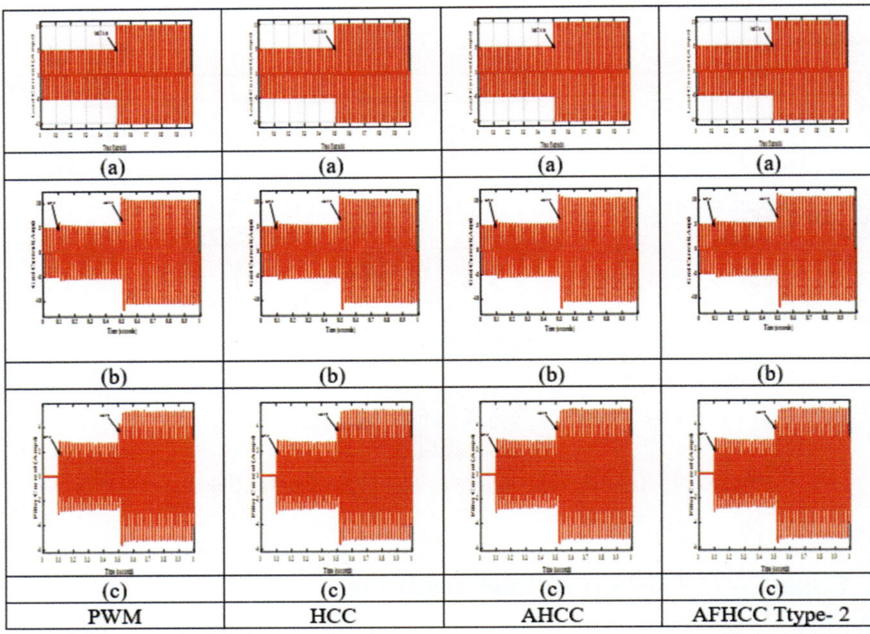

Figure 4.10 Simulation results of (a) grid current before compensation, (b) compensation currents and (c) grid current after compensation.

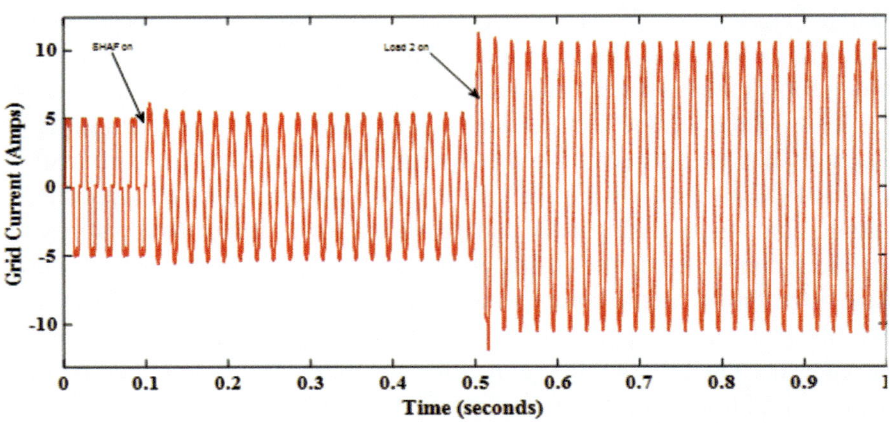

Figure 4.11 Double-tuned SHAF grid currents with PI DC-link voltage controller and HCC switching scheme.

became sinusoidal at 0.1 sec. The double-tuned SHAF is able to compensate the load current harmonics even under dynamic loading condition at 0.5 sec.

The SAF compensation current with PI DC-link voltage controller and HCC switching scheme is shown in Figure 4.12. The SAF performance is not affected with simultaneous operation of tuned passive filters.

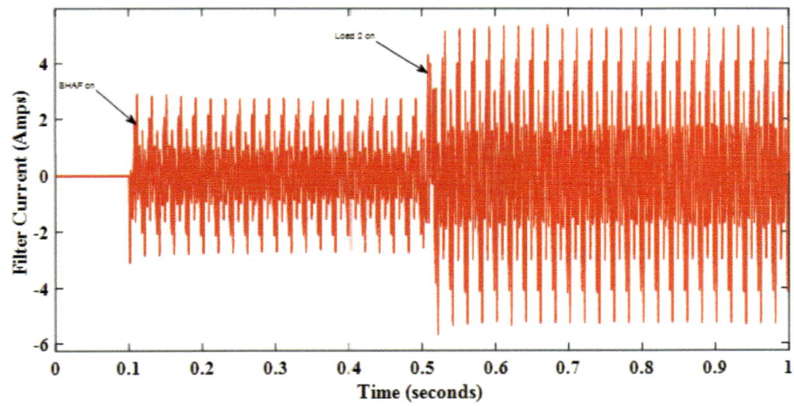

Figure 4.12 Double-tuned SHAF compensating currents with PI DC-link voltage controller and HCC switching scheme.

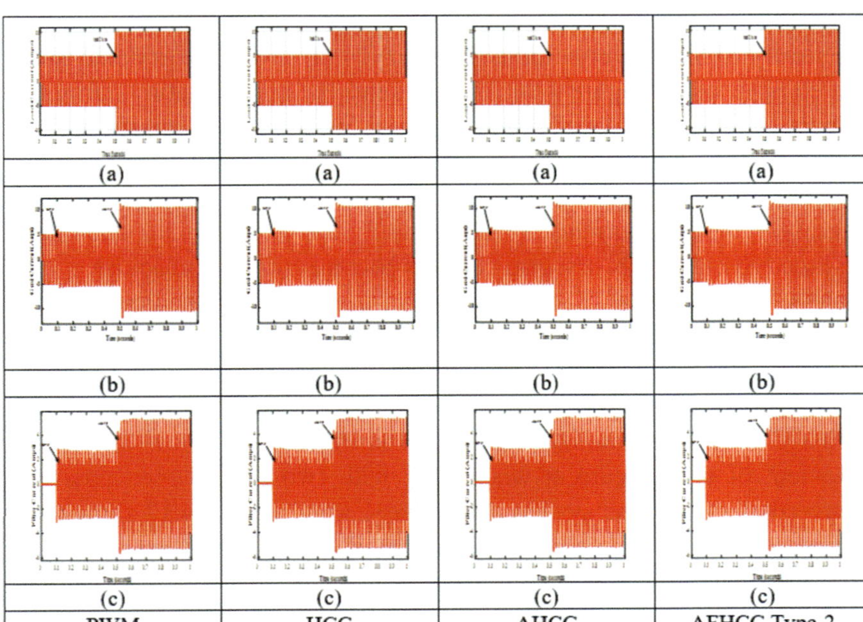

Figure 4.13 Simulation results of (a) grid current before compensation, (b) compensation currents, and (c) grid current after compensation.

4.2.2 Case-2 – Type-1 fuzzy logic DC-link voltage controller

Type-1 fuzzy logic DC-link voltage controller with PWM/HCC/AHCC/AFHCCtype-1/AFHCC type-2 pulse generators are simulated under uncontrolled bridge rectifiers with RL load. Figure 4.13 is achieved by using PWM/HCC/AHCC/AFHCC type-2 under uncontrolled diode bridge rectifier loads with steady-state and transient state conditions. Figure 4.13 (a) shows the source currents without compensation. Figure 4.13 (b) shows the SAF compensating currents and Figure 4.13 (c) shows the grid currents after compensation. To mitigate the harmonic currents, the double-tuned SHAF (when it is ON) has to provide the desired compensating currents as shown in Figure 4.13 (b). The grid currents after compensation are shown in Figure 4.13 (c), the double-tuned SHAF is ON, the source currents become sinusoidal.

4.2.3 Case-3 – Type-2 fuzzy logic DC-link voltage controller

Type-2 fuzzy logic DC-link voltage controllers with PWM/HCC/AHCC /AFHCC type-1/AFHCC type-2 are simulated under uncontrolled diode bridge rectifiers with RL load. Figure 4.14 is achieved using PWM/HCC/AHCC/AFHCC type-1 under uncontrolled diode bridge rectifier loads with steady- and transient-state conditions. Figure 4.14 (a) shows the grid currents without compensation. Figure 4.14 (b) represents the compensating currents and Figure 4.14 (c) shows the grid currents with SHAF compensation. To compensate the harmonic currents, the double-tuned SHAF has to give the desired compensating currents as represented in Figure 4.14 (b). After compensation, the grid currents are shown in Figure 4.14 (c), i.e., when the double-tuned SHAF is ON, the grid currents become mostly sinusoidal.

Table 4.1 and Figure 4.15 show the THD measurement of source current with only SAF. The performance of SAF is analyzed with unit vector control as a reference current controller, DC-link controller (PI, type-1 fuzzy and type-2 fuzzy) and pulse generators (PWM, HCC, AHCC, AFHCC type-1 and AFHCC type-2). Similarly, Table 4.2 and Figure 4.16 show the THD measurement of source current when SAF with two single-tuned passive filters are connected to the system and Table 4.3 and Figure 4.17 show the THD measurement of source current when SAF with double-tuned passive filter is connected.

Tables 4.4–4.9 give the information regarding the real and reactive power requirements when connected to SAF, single- and double-tuned SHAF to the non-linear system under steady- and transient-state conditions.

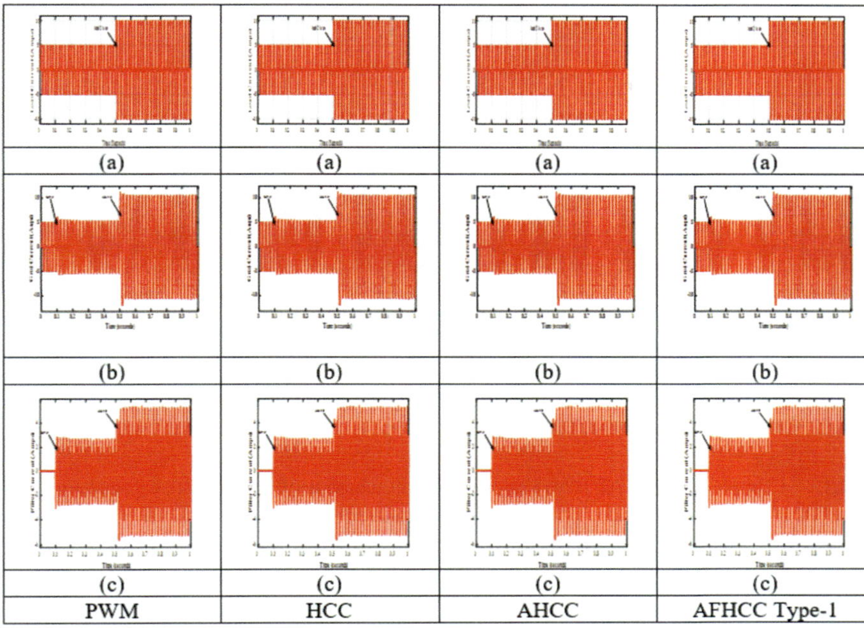

Figure 4.14 Simulation results of (a) grid current before compensation, (b) compensation currents, and (c) grid current after compensation.

Table 4.1 THD measurements of SAF with unit vector control.

Load	Harmonic controller	DC-link controller	Condition	Type of pulse generator				
				PWM	HCC	AHCC	AFHCC type-1	AF-HCC type-2
Diode bridge rectifier with RL load	Unit vector	PI	Steady State	4.96	4.2	4.1	4	3.9
			Transient	4.9	4.12	3.99	3.91	3.8
		Type-1 fuzzy	Steady State	4.8	4.01	3.96	3.83	3.77
			Transient	4.74	3.98	3.92	3.8	3.71
		Type-2 fuzzy	Steady state	4.7	3.96	3.9	3.75	3.67
			Transient	4.61	3.9	3.85	3.72	3.6

Figure 4.18 shows the reactive power balance and real power demand under steady and transient state conditions using SHAF. In Figure 4.18, the SHAF utilizes type-2 fuzzy logic DC-link voltage controller, AFHCC type-2 pulse generator, and double-tuned passive filter. In a similar manner, the oscillating component of reactive power, real power, and power factor has been eliminated by all the proposed SAF, single- and double-tuned SHAF.

Figure 4.19 shows the DC-link voltage across the VSI capacitor bank of SHAF. In Figure 4.19, the SHAF utilizes type-2 fuzzy logic DC-link voltage

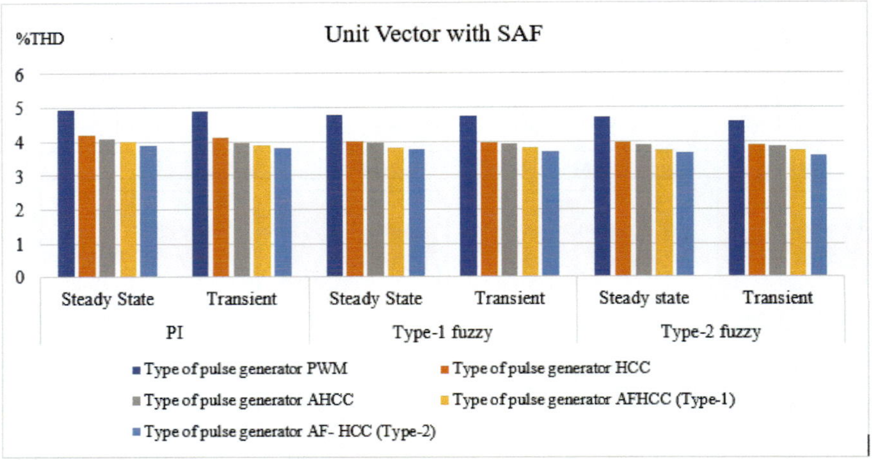

Figure 4.15 Comparison of SAF with various DC voltage controllers and pulse generators.

Table 4.2 THD measurements of single-tuned SHAF with unit vector control.

Load	Harmonic controller	DC-link controller	Condition	Type of pulse generator				
				PWM	HCC	AHCC	AFHCC (type-1)	AF-HCC (type-2)
Diode bridge rectifier with RL load	Unit vector	PI	Steady State	4.23	3.82	3.72	3.6	3.51
			Transient	4.12	3.78	3.65	3.53	3.42
		Type-1 fuzzy	Steady State	4.1	3.6	3.59	3.51	3.4
			Transient	4.0	3.53	3.51	3.43	3.35
		Type-2 fuzzy	Steady state	3.9	3.4	3.45	3.4	3.22
			Transient	3.85	3.36	3.39	3.35	3.13

controller, AFHCC type-2 pulse generator, and double-tuned passive filter. In a similar manner, all the DC-link control techniques of SAF are able to maintain the constant DC-link voltage across VSI capacitor bank under steady- and transient-state conditions.

THD analysis: The harmonics against magnitude of the grid current is plotted with double-tuned SHAF under uncontrolled bridge rectifier with RL load. The simulation results in Figures 4.20 and 4.21 are for type-2 fuzzy DC capacitor voltage controller and AFHCC type-2 pulse generator-based double-tuned SHAF under steady and transient state conditions with diode

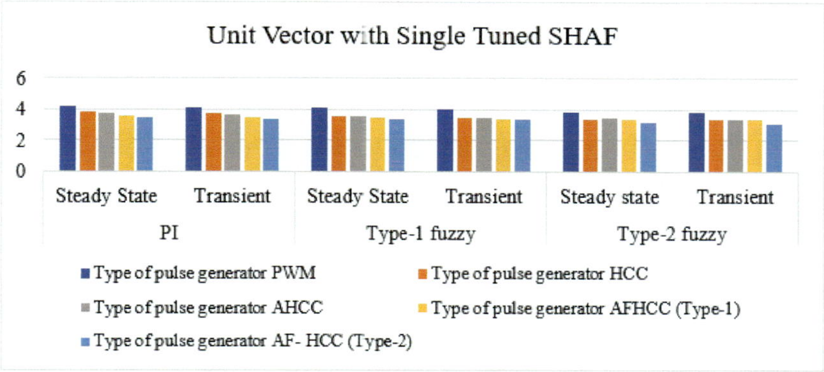

Figure 4.16 Comparison of single-tuned SHAF with various DC voltage controllers and pulse generators.

Table 4.3 THD measurements of double-tuned SHAF with unit vector control.

Load	Harmonic controller	DC-link controller	Condition	Type of pulse generator				
				PWM	HCC	AHCC	AFHCC (type-1)	AF-HCC (type-2)
Diode bridge rectifier with RL load	Unit vector	PI	Steady State	4.13	3.79	3.62	3.55	3.42
			Transient	4.1	3.71	3.54	3.46	3.38
		Type-1 fuzzy	Steady State	3.95	3.62	3.42	3.33	3.36
			Transient	3.9	3.52	3.4	3.38	3.35
		Type-2 fuzzy	Steady state	3.82	3.5	3.34	3.21	**3.33**
			Transient	3.9	3.6	3.53	3.5	**3.45**

bridge rectifier. It has been observed that the grid current without SHAF is having harmonic components whereas with double-tuned SHAF, the THD fell from 26.12% to 3.33% in steady state and from 26.21% to 3.45% in transient state respectively as represented in Figures 4.20 and 4.21.

Real and Reactive power analysis: The SAF, single-tuned SHAF, and double-tuned SHAF improve the power factor in a distribution network. Figure 4.22 shows the waveform of power factor correction for a phase voltage and current when SHAF is switched on at 0.1 sec. The real and reactive powers for SAF, single- and double-tuned SHAF are calculated as given in Tables 4.4 to 4.9. This result indicates that, the proposed SHAF is diminishing reactive power and therefore improving the power factor.

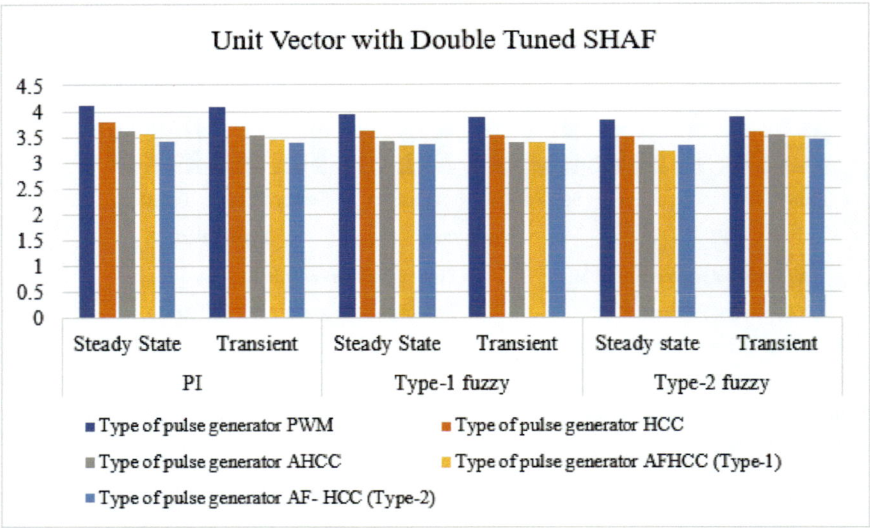

Figure 4.17 Comparison of double-tuned SHAF with various DC voltage controllers and pulse generators.

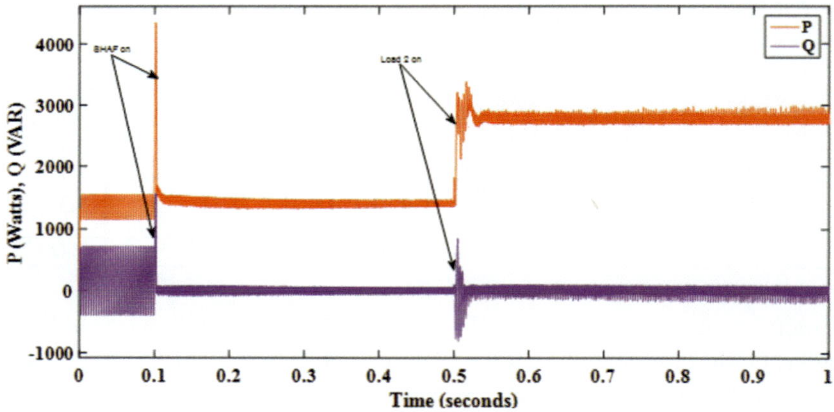

Figure 4.18 Real power and reactive power demands.

Remarks: The PI, type-1 fuzzy and type-2 fuzzy DC capacitor voltages in association with the various pulse generators for SHAF has been simulated and performance is evaluated. The type-2 AFHCC reduces the switching power losses and improves the SHAF performance compared with the PWM, HCC, AHCC, and type-1 AFHCC. Similarly, type-2 fuzzy is giving better

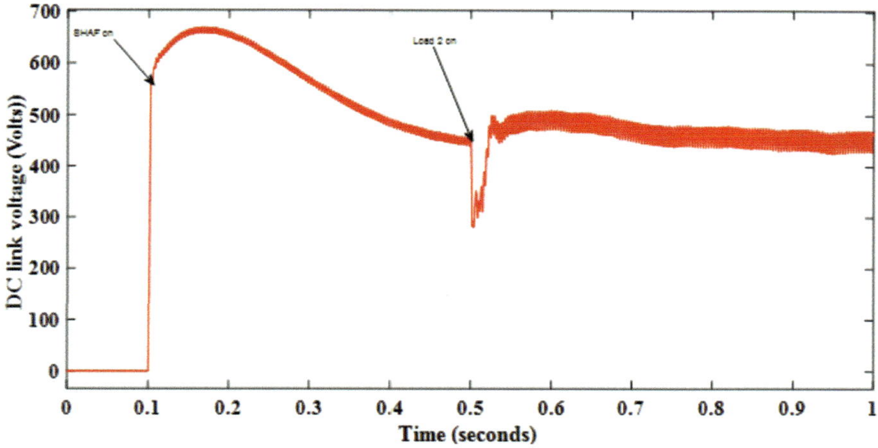

Figure 4.19 DC-link voltage across the VSI capacitor.

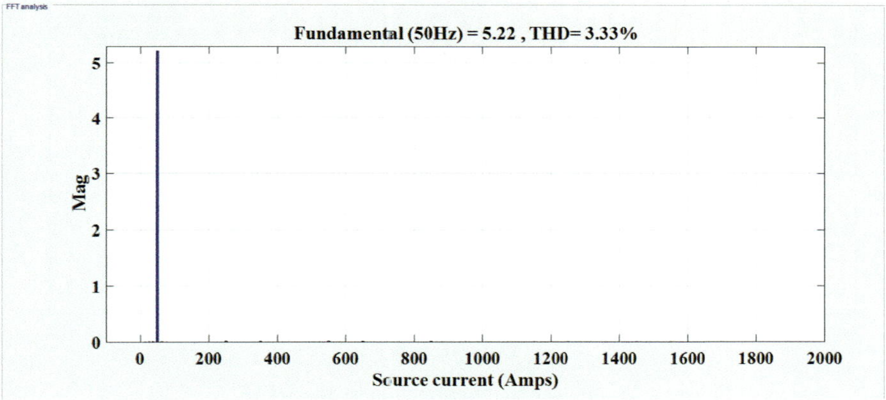

Figure 4.20 THD analysis of double-tuned SHAF under steady-state conditions.

performance than conventional PI or type-1 fuzzy in terms of VDC-settling time, harmonic current and reactive-power compensation. The THD of the grid current is less than 5% of all the cases which are following IEE-519-2022 standards.

4.3 Fryze Power Theory

The Fryze power theory with different PWM pulse generators, DC capacitor voltage regulators, and tuned passive filters is performed by using MATLAB/Simulink tools. The model system is having three-phase VSI, coupling

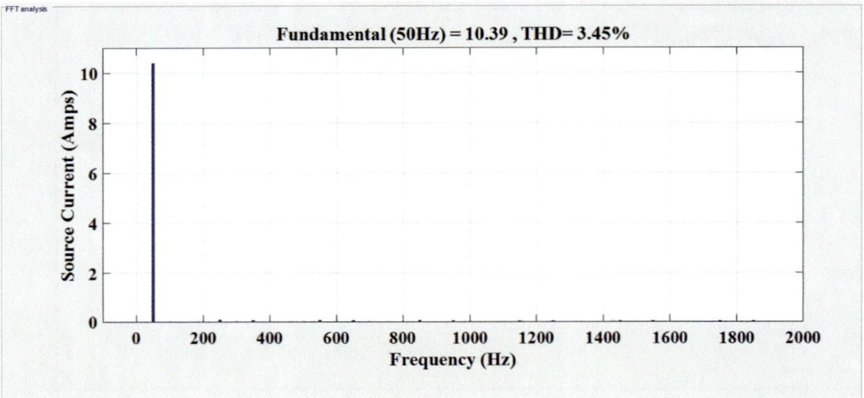

Figure 4.21 THD analysis of double-tuned SHAF under transient-state conditions.

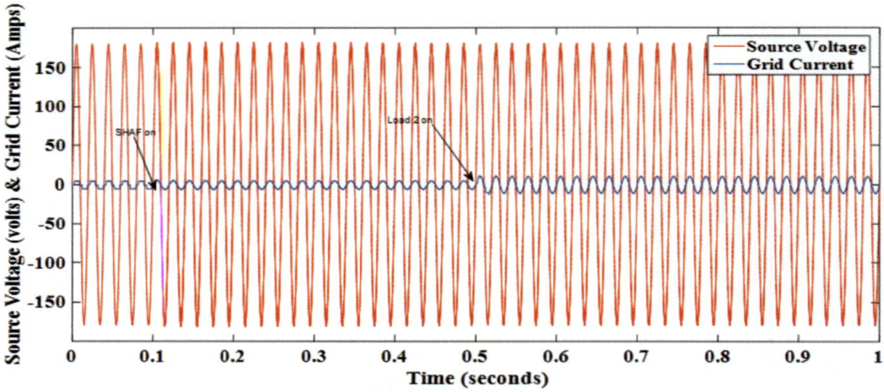

Figure 4.22 Power factor correction by the SHAF.

inductor, time-domain reference current extraction based on Fryze power theory, pulse generator, and tuned passive filters as represented in Figure 4.23. It is used to pull out the desired reference current from the polluted load current. The reference current is related with original compensation current to produce the trigger pulses for SAF using PWM/HCC/AHCC/AFHCC type-1/AFHCC type-2 controllers. The SAF generates the anti-harmonic current to cancel out current harmonics and balance reactive power in the distributed network. The whole system is tested under steady- and transient-state conditions. Further, the tuned passive filters like single-tuned and double-tuned type combines with the SAF which gives better performance.

Table 4.4 Real and reactive powers of SAF with unit vector control under steady state.

Load	Reference controller	Without SAF	DC-link controller	With SAF (steady state)				
				PWM	HCC	AHCC	AFHCC type-1	AFHCC type-2
Diode bridge rectifier with RL load	Unit vector	$P =$ 1.5 KW $Q =$ 720 VAR	PI	$P =$ 1.46 KW $Q =$ 60 VAR	$P =$ 1.47 KW $Q =$ 29 VAR	$P =$ 1.45 KW $Q =$ 49 VAR	$P =$ 1.49 KW $Q =$ 25.5 VAR	$P =$ 1.48 KW $Q =$ 43 VAR
			Type-1 fuzzy	$P =$ 1.45 KW $Q =$ 100 VAR	$P =$ 1.46 KW $Q =$ 29 VAR	$P =$ 1.48 KW $Q =$ 32 AR	$P =$ 1.48 KW $Q =$ 50 VAR	$P =$ 1.48 KW $Q =$ 50 VAR
			Type-2 fuzzy	$P =$ 1.45 KW $Q =$ 100 VAR	$P =$ 1.46 KW $Q =$ 29 VAR	$P =$ 1.48 KW $Q =$ 32 VAR	$P =$ 1.47 KW $Q =$ 50 VAR	$P =$ 1.49 KW $Q =$ 50 VAR

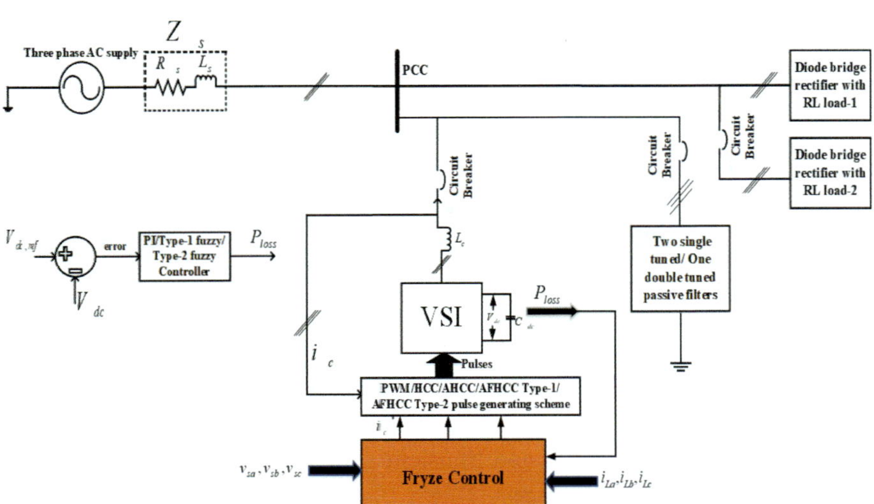

Figure 4.23 Fryze power theory-based SHAF.

Table 4.5 Real and reactive powers of single-tuned SHAF with unit vector control under steady state.

Load	Reference controller	Without SAF	DC-link controller	With single-tuned SHAF (steady state)				
				PWM	HCC	AHCC	AFHCC type-1	AFHCC type-2
Diode bridge rectifier with RL load	Unit vector	$P=$ 1.5 KW $Q=$ 720 VAR	PI	$P=$ 1.46 KW $Q=$ 60 VAR	$P=$ 1.49 KW $Q=$ 29 VAR	$P=$ 1.48 KW $Q=$ 49 VAR	$P=$ 1.49 KW $Q=$ 25.5 VAR	$P=$ 1.49 KW $Q=$ 43 VAR
			Type-1 fuzzy	$P=$ 1.45 KW $Q=$ 100 VAR	$P=$ 1.46 KW $Q=$ 29 VAR	$P=$ 1.48 KW $Q=$ 32 VAR	$P=$ 1.49 KW $Q=$ 50 VAR	$P=$ 1.48 KW $Q=$ 50 VAR
			Type-2 fuzzy	$P=$ 1.45 KW $Q=$ 100 VAR	$P=$ 1.46 KW $Q=$ 29 VAR	$P=$ 1.48 KW $Q=$ 32 VAR	$P=$ 1.47 KW $Q=$ 50 VAR	$P=$ 1.49 KW $Q=$ 50 VAR

4.3.1 Case-1 – PI DC-link voltage controller

PI DC-link voltage controller with different pulse generators is simulated for SHAF under balanced uncontrolled diode bridge rectifiers with RL-load conditions. The waveforms in Figure 4.24 are concentrated on PI DC-link voltage controller with PWM/HCC/AHCC/AFHCC type-1/AFHCC type-2 techniques respectively. Figure 4.24 (a) shows the load current without SHAF. It is observed that the load current is having power frequency as well as harmonics due to the non-linear load (diode bridge rectifier load). The SHAF when switched on at 0.1 sec provides the desired compensating current to cancel out the harmonic current at PCC as shown in Figure 4.24 (b). By injecting equal but opposite harmonic components, the harmonic compensation has been achieved. Figure 4.24 (c) shows the grid current after harmonic compensation. These waveforms are achieved under uncontrolled bridge rectifier load with steady- and transient-state conditions and these are for A-phase alone. The THD measurement of source/load current without

Table 4.6 Real and reactive powers of double-tuned SHAF with unit vector control under steady state.

Load	Reference controller	Without SAF	DC-link controller	With double tuned SHAF (steady state)				
				PWM	HCC	AHCC	AFHCC type-1	AFHCC type-2
Diode bridge rectifier with RL load	Unit vector	P = 1.5 KW Q = 720 VAR	PI	$P =$ 1.46 KW $Q =$ 60 VAR	$P =$ 1.45 KW $Q =$ 29 VAR	$P =$ 1.46 KW $Q =$ 49 VAR	$P =$ 1.49 KW $Q =$ 25.5 VAR	$P =$ 1.46 KW $Q =$ 43 VAR
			Type-1 fuzzy	$P =$ 1.45 KW $Q =$ 100 VAR	$P =$ 1.46 KW $Q =$ 29 VAR	$P =$ 1.48 KW $Q =$ 32 VAR	$P =$ 1.55 KW $Q =$ 50 VAR	$P =$ 1.48 KW $Q =$ 50 VAR
			Type-2 fuzzy	$P =$ 1.45 KW $Q =$ 100 VAR	$P =$ 1.46 KW $Q =$ 29 VAR	$P =$ 1.48 KW $Q =$ 32 VAR	$P =$ 1.55 KW $Q =$ 50 VAR	$P =$ 1.49 KW $Q =$ 50 VAR

filter under steady-state condition is 26.12% and under transient condition is 26.01%.

4.3.2 Case-2 – Type-1 fuzzy logic DC-link voltage controller

Type-1 fuzzy logic DC-link voltage controllers with PWM/HCC/AHCC/AFHCC type-1/AFHCC type-2 are simulated under uncontrolled bridge rectifier with RL load. Figure 4.25 is achieved by using PWM/HCC/AHCC/AFHCC type-2 under uncontrolled diode bridge rectifier load with steady and transient states. Figure 4.25 (a) shows the source currents without compensation. Figure 4.25 (b) shows the compensating currents and Figure 4.25 (c) shows the grid currents after compensation. To mitigate the harmonic currents, the double-tuned SHAF (when it is ON) has to provide the desired compensating currents as shown in Figure 4.25 (b). The grid currents after compensation are displayed in Figure 4.25 (c). When the double-tuned SHAF is activated, the source currents become sinusoidal.

Table 4.7 Real and reactive powers of SAF with unit vector control under transient state.

Load	Reference controller	Without SAF	DC-link controller	With SAF (transient state)				
				PWM	HCC	AHCC	AFHCC type-1	AFHCC type-2
Diode bridge rectifier with RL load	Unit vector	P = 3 KW Q = 1540 VAR	PI	$P =$ 2.8 KW $Q =$ 90 VAR	$P =$ 2.81 KW $Q =$ 92 VAR	$P =$ 2.78 KW $Q =$ 91 VAR	$P =$ 2.75 KW $Q =$ 92 VAR	$P =$ 2.79 KW $Q =$ 91 VAR
			Type-1 fuzzy	$P =$ 2.8 KW $Q =$ 90 VAR	$P =$ 2.81 KW $Q =$ 92 VAR	$P =$ 2.78 KW $Q =$ 91 VAR	$P =$ 2.75 KW $Q =$ 92 VAR	$P =$ 2.79 KW $Q =$ 91 VAR
			Type-2 fuzzy	$P =$ 2.8 KW $Q =$ 90 VAR	$P =$ 2.81 KW $Q =$ 92 VAR	$P =$ 2.78 KW $Q =$ 91 VAR	$P =$ 2.75 KW $Q =$ 92 VAR	$P =$ 2.79 KW $Q =$ 91 VAR

Figure 4.26 shows the grid current of phase A, when SAF with type-1 fuzzy logic DC-link voltage controller and AHCC switching scheme is connected to the double-tuned passive filter at 0.1 sec. It has been noticed that, the current became sinusoidal at 0.1 sec. The double-tuned SHAF is also able to compensate under dynamic loading condition at 0.5 sec.

The SAF compensation currents with type-1 fuzzy logic DC-link voltage controller and AHCC switching scheme, and double-tuned passive filter is as shown in Figure 4.27.

4.3.3 Case-3 – Type-2 fuzzy logic DC-link voltage controller

Type-2 fuzzy logic pulse generators with PWM/HCC/AHCC/AFHCC type-2/AFHCC type-2 are simulated under uncontrolled diode bridge rectifier with RL load. Figure 4.28 is achieved using PWM/HCC/AHCC/AFHCC type-2 under uncontrolled diode bridge rectifier load with steady and transient states. Figure 4.28 (a) shows the grid currents without compensation. Figure 4.28 (b) represents the compensating currents and Figure 4.28 (c) shows the grid

Table 4.8 Real and reactive powers of single-tuned SHAF with unit vector control under transient state.

Load	Reference controller	Without SAF	DC-link controller	With single tuned SHAF (transient state)				
				PWM	HCC	AHCC	AFHCC type-1	AFHCC type-2
Diode bridge rectifier with RL load	Unit vector	$P = 3$ KW $Q = 1540$ VAR	PI	$P = 2.8$ KW $Q = 90$ VAR	$P = 2.8$ KW $Q = 92$ VAR	$P = 2.78$ KW $Q = 90$ VAR	$P = 2.7$ KW $Q = 92$ VAR	$P = 2.75$ KW $Q = 91$ VAR
			Type-1 fuzzy	$P = 2.8$ KW $Q = 90$ VAR	$P = 2.81$ KW $Q = 92$ VAR	$P = 2.76$ KW $Q = 91$ VAR	$P = 2.75$ KW $Q = 92$ VAR	$P = 2.78$ KW $Q = 91$ VAR
			Type-2 fuzzy	$P = 2.8$ KW $Q = 90$ VAR	$P = 2.81$ KW $Q = 92$ VAR	$P = 2.78$ KW $Q = 91$ VAR	$P = 2.75$ KW $Q = 92$ VAR	$P = 2.79$ KW $Q = 91$ VAR

currents with SHAF compensation. To compensate the harmonic currents, the double-tuned SHAF has to give the desired compensating currents as represented in Figure 4.28 (b). After compensation, the grid currents are shown in Figure 4.28 (c), i.e., when the double-tuned SHAF is ON, the grid currents become mostly sinusoidal.

THD analysis: The harmonics against magnitude of the grid current is plotted with double-tuned SHAF under uncontrolled bridge rectifier with RL load. The simulation results are shown in Figure 4.29 is for type-2 fuzzy DC capacitor voltage controller and AFHCC type-2 pulse generator-based double-tuned SHAF under steady- and transient-state conditions with diode bridge rectifier. It has been observed that the grid current without SHAF is having harmonic components whereas with double-tuned SHAF, the THD fell from 26.12% to 3.91% in steady-state and 26.21% to 3.27% in transient-state respectively as represented in Figure 4.29.

Table 4.10 and Figure 4.30 show the THD measurement of source current with only SAF. The performance of SAF is analyzed with Fryze power

Table 4.9 Real and reactive power requirements of double-tuned SHAF with unit control under transient state.

Load	Reference controller	Without SAF	DC-link controller	With double tuned SHAF (transient state)				
				PWM	HCC	AHCC	AFHCC type-1	AFHCC type-2
Diode bridge rectifier with RL load	Unit vector	$P = 3$ KW $Q =$ 1540 VAR	PI	$P =$ 2.88 KW $Q =$ 91 VAR	$P =$ 2.78 KW $Q =$ 90 VAR	$P =$ 2.76 KW $Q =$ 92 VAR	$P =$ 2.78 KW $Q =$ 92 VAR	$P =$ 2.74 KW $Q =$ 91 VAR
			Type-1 fuzzy	$P =$ 2.79 KW $Q =$ 90 VAR	$P =$ 2.81 KW $Q =$ 92 VAR	$P =$ 2.78 KW $Q =$ 91 VAR	$P =$ 2.75 KW $Q =$ 92 VAR	$P =$ 2.79 KW $Q =$ 91 VAR
			Type-2 fuzzy	$P =$ 2.81 KW $Q =$ 91 VAR	$P =$ 2.81 KW $Q =$ 92 VAR	$P =$ 2.76 KW $Q =$ 92 VAR	$P =$ 2.78 KW $Q =$ 91 VAR	$P =$ 2.74 KW $Q =$ 91 VAR

control as a reference current controller, DC-link controller (PI, type-1 fuzzy, and type-2 fuzzy), and pulse generators (PWM, HCC, AHCC, AFHCC type-1, and AFHCC type-2). Similarly, Table 4.11 and Figure 4.31 show the THD measurement of source current when SAF with two single-tuned passive filters are connected to the system and Table 4.12 and Figure 4.32 show the THD measurement of source current when SAF with double-tuned passive filter is connected.

Real and Reactive power analysis: The SAF, single-tuned SHAF, and double-tuned SHAF improve the power factor in distribution network. Figure 4.33 shows the waveform of unity power factor when SHAF is switched on at 0.1 sec. The real and reactive powers are calculated as given in Tables 4.13–4.15. This result indicates that, the SHAF is diminishing reactive power and therefore improving the power factor.

Tables 4.13 to 4.15 give the information regarding the real and reactive power requirements when connected to SAF, single-tuned SHAF, and double-tuned SHAF to the non-linear system under steady-state conditions.

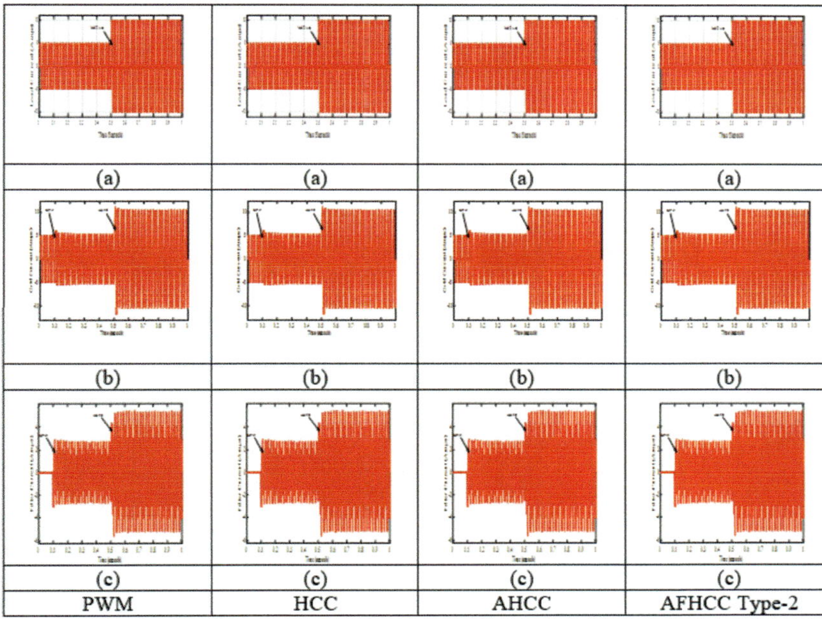

Figure 4.24 Simulation waveforms of (a) source current before compensation, (b) compensation currents, and (c) source current after compensation.

Table 4.10 THD measurements of SAF with Fryze control.

Load	Harmonic controller	DC-link controller	Condition	Type of pulse generator				
				PWM	**HCC**	**AHCC**	**AFHCC (type-1)**	AF-HCC (type-2)
Diode bridge rectifier with RL load	Fryze	PI	Steady State	4.82	4.75	4.65	4.55	4.5
			Transient	4.8	4.7	4.61	4.59	4.52
		Type-1 fuzzy	Steady State	4.75	4.67	4.6	4.52	4.45
			Transient	4.71	4.65	4.59	4.49	4.4
		Type-2 fuzzy	Steady state	4.69	4.53	4.45	4.41	4.38
			Transient	4.6	4.51	4.44	4.38	4.31

The DC-link voltage across the VSI capacitor bank of SHAF is as shown in Figure 4.34. In Figure 4.34, the SHAF utilizes type-2 fuzzy logic DC-link voltage controller, AFHCC type-2 pulse generator, and double-tuned passive

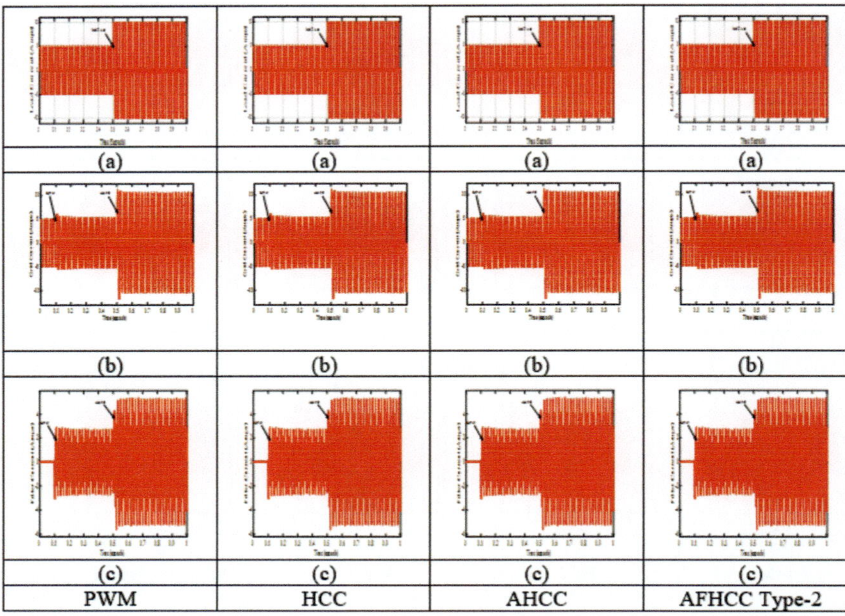

Figure 4.25 Simulation waveforms of (a) source current before compensation, (b) compensation currents, and (c) source current after compensation.

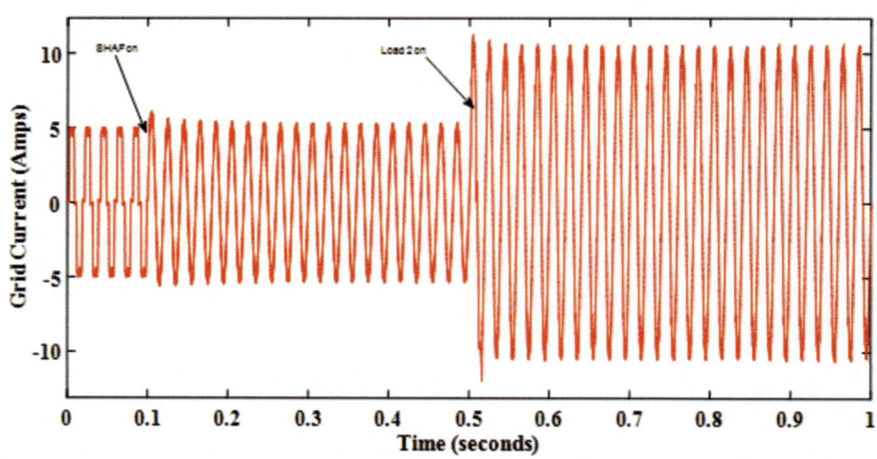

Figure 4.26 Double-tuned SHAF grid currents with PI DC-link voltage controller and HCC switching scheme.

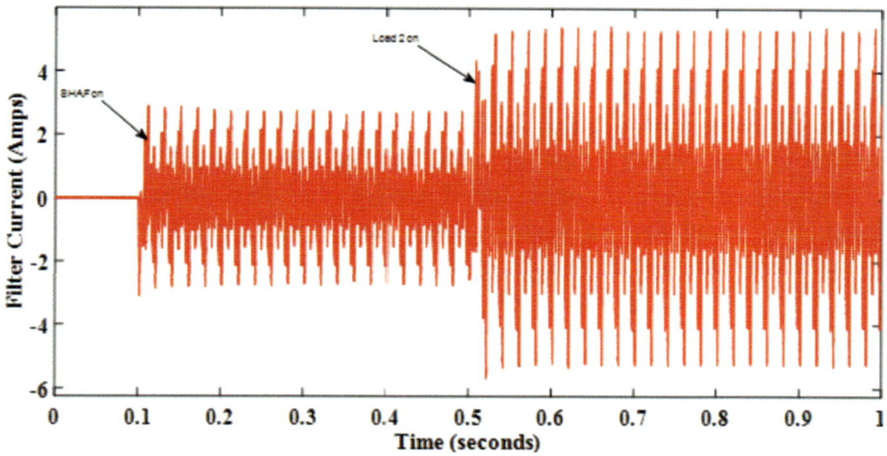

Figure 4.27 Double-tuned SHAF compensating currents with PI DC-link voltage controller and HCC switching scheme.

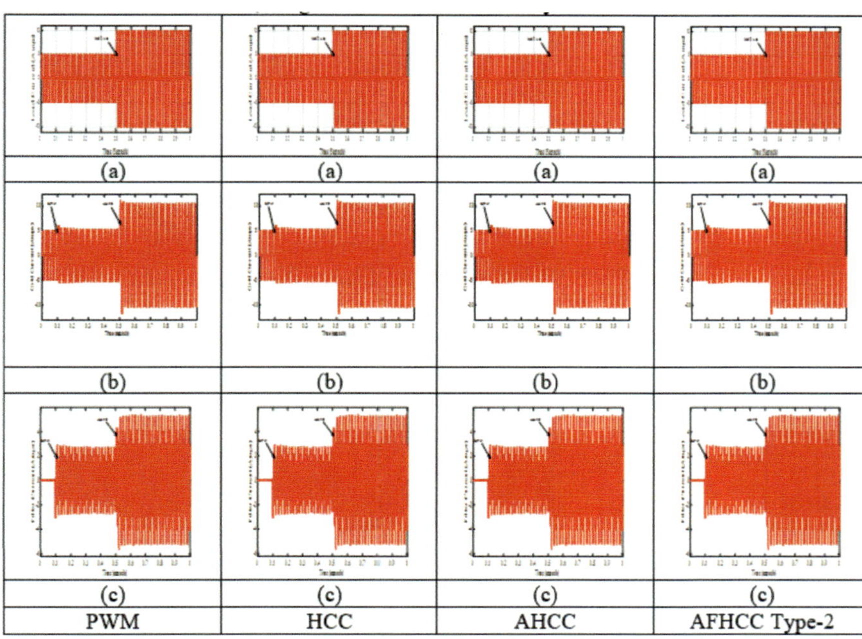

Figure 4.28 Simulation waveforms of (a) source current before compensation, (b) compensation currents, and (c) source current after compensation.

filter. In a similar manner, all the DC-link control techniques of SAF are able to maintain the constant DC-link voltage across VSI capacitor bank under steady- and transient-state conditions.

Figure 4.35 shows the reactive power balance and real power demand under steady- and transient-state condition using SHAF. In Figure 4.35, the SHAF utilizes type-2 fuzzy logic DC-link voltage controller, AFHCC type-2 pulse generator, and double-tuned passive filter. In a similar manner, the oscillating component of reactive power, real power, and power factor has been eliminated by all the proposed SAF, single-tuned, and double-tuned SHAF. Figure 4.36 illustrates the power factor correction when the double-tuned SHAF is connected at 0.1 seconds.

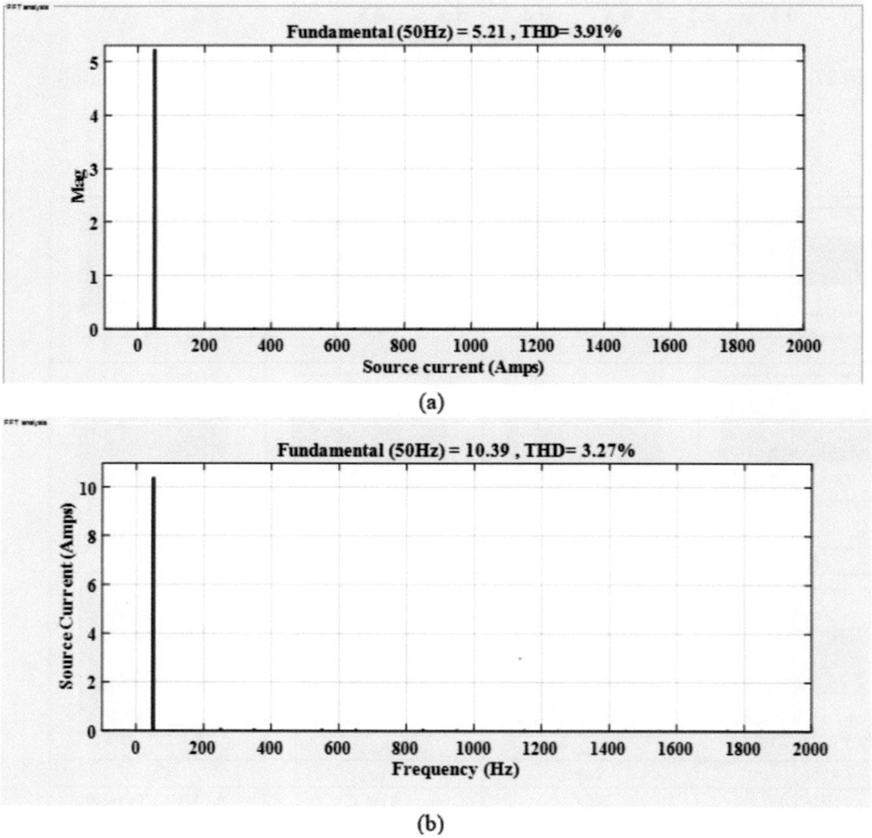

Figure 4.29 THD analysis of double-tuned SHAF under (a) steady-state (b) transient-state conditions.

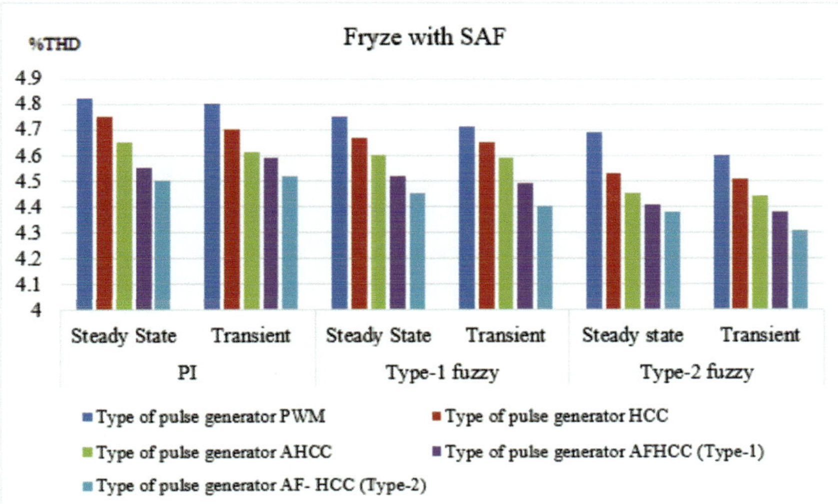

Figure 4.30 Comparison of SAF with various DC voltage controllers and pulse generators.

Remarks: The Fryze power theory with various PWM pulse generators-based SHAF is simulated and compensation principle is evaluated. The AFHCC type-2 reduces the switching power losses and improves the SHAF performance than other techniques. From these results, it is proved that the THD of the grid current is following harmonic standards IEEE 519-2014.

Table 4.11 THD measurements of single-tuned SHAF with Fryze control.

Load	**Harmonic controller**	**DC-link controller**	**Condition**	**Type of pulse generator**				
				PWM	**HCC**	**AHCC**	**AFHCC (type-1)**	AF-HCC (type-2)
Diode bridge rectifier with RL load	Fryze	PI	Steady State	4.55	4.32	4.25	4.2	4.15
			Transient	4.51	4.28	4.1	3.98	3.9
		Type-1 fuzzy	Steady State	4.45	4.4	4.35	4.3	4.28
			Transient	4.42	4.38	4.31	4.28	4.25
		Type-2 fuzzy	Steady state	4.32	4.28	4.21	4.15	4.1
			Transient	4.28	4.25	4.18	4.1	3.98

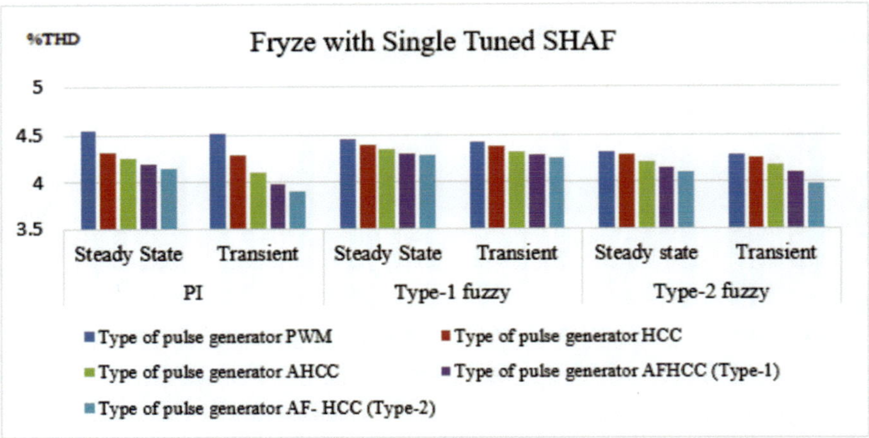

Figure 4.31 Comparison of single-tuned SHAF with various DC voltage controllers and pulse generators.

Table 4.12 THD measurements of double-tuned SHAF with Fryze control.

Load	**Harmonic controller**	**DC-link controller**	**Condition**	**Type of pulse generator**				
				PWM	**HCC**	**AHCC**	**AFHCC (type-1)**	**AF-HCC (type-2)**
Diode bridge rectifier with RL load	Fryze	PI	Steady State	4.28	4.25	4.2	4.15	4.1
			Transient	4.25	4.21	4.15	4.1	4.01
		Type-1 fuzzy	Steady State	4.15	4.1	3.98	3.9	3.85
			Transient	4.12	4.05	3.9	3.85	3.81
		Type-2 fuzzy	Steady state	4.1	4.01	3.98	3.95	**3.91**
			Transient	4.12	3.92	3.86	3.35	**3.27**

4.4 Instantaneous Real and Reactive Power (p-q) Theory

The p-q theory with different PWM pulse generators, DC capacitor voltage regulators, and tuned passive filters is performed by using MATLAB/Simulink tools. The model system is having three-phase VSI, coupling inductor, time-domain reference current extraction controller based on p-q theory, pulse generator, and tuned passive filters as represented in Figure 4.37. It is used to pull out the desired reference current from the polluted load

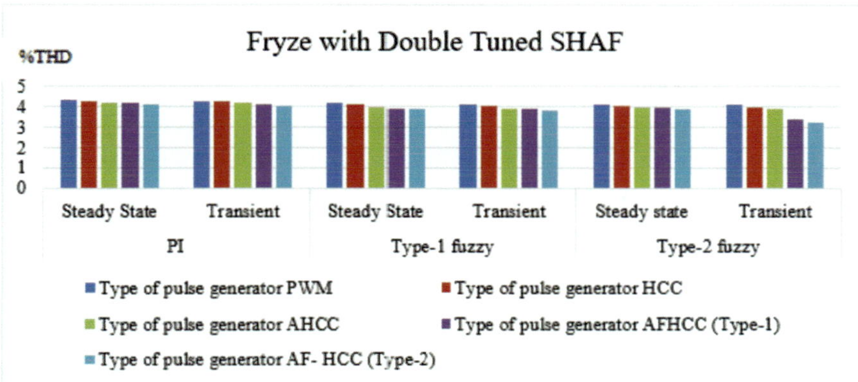

Figure 4.32 Comparison of double-tuned SHAF with various DC voltage controllers and pulse generators.

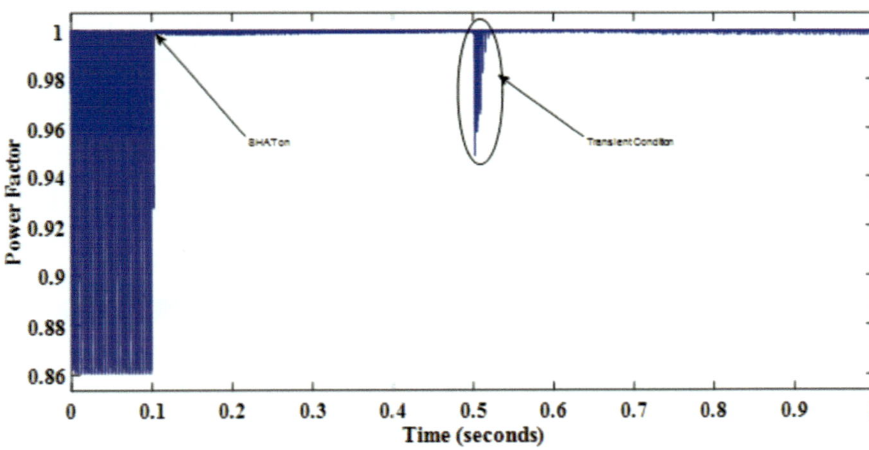

Figure 4.33 Unity power factor by the SHAF.

current. The reference current is related with original compensation current to produce the trigger pulses for SAF using PWM/HCC/AHCC/AFHCC type-1/AFHCC type-2 controllers. The SAF generates the anti-harmonic current to cancel out current harmonics and balance reactive power in the distributed network. The whole system is tested under steady- and transient-state conditions. Further, the tuned passive filters like single-tuned and double-tuned type combine with the SAF and gives better performance.

Table 4.13 Real and reactive powers of SAF with Fryze control under steady state.

Load	Reference controller	Without SAF	DC-link controller	With SAF (steady state)				
				PWM	HCC	AHCC	AFHCC type-1	AFHCC type-2
Diode bridge rectifier with RL load	Fryze	P = 1.5 KW Q = 720 VAR	PI	$P =$ 1.45 KW $Q =$ 59 VAR	$P =$ 1.43 KW $Q =$ 29 VAR	$P =$ 1.45 KW $Q =$ 49 VAR	$P =$ 1.49 KW $Q =$ 25.5 VAR	$P =$ 1.45 KW $Q =$ 42 VAR
			Type-1 fuzzy	$P =$ 1.45 KW $Q =$ 100 VAR	$P =$ 1.46 KW $Q =$ 29 VAR	$P =$ 1.48 KW $Q =$ 32 VAR	$P =$ 1.45 KW $Q =$ 50 VAR	$P =$ 1.48 KW $Q =$ 50 VAR
			Type-2 fuzzy	$P =$ 1.45 KW $Q =$ 100 VAR	$P =$ 1.46 KW $Q =$ 29 VAR	$P =$ 1.48 KW $Q =$ 32 VAR	$P =$ 1.45 KW $Q =$ 50 VAR	$P =$ 1.49 KW $Q =$ 49 VAR

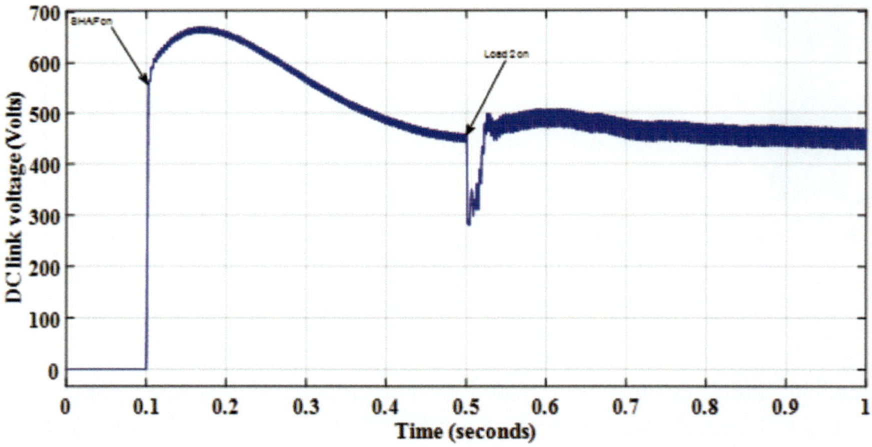

Figure 4.34 DC-link voltage across the VSI capacitor.

4.4.1 Case-1 – PI DC-link voltage controller

The PI DC-link voltage controller with different pulse generators is simulated for SHAF under balanced uncontrolled diode bridge rectifiers with RL-load

Table 4.14 Real and reactive powers of single-tuned SHAF with Fryze control under steady state.

Load	Reference controller	Without SAF	DC-link controller	With single-tuned SHAF (steady state)				
				PWM	HCC	AHCC	AFHCC type-1	AFHCC type-2
Diode bridge rectifier with RL load	Fryze	P = 1.5 KW Q = 720 VAR	PI	$P =$ 1.46 KW $Q =$ 59 VAR	$P =$ 1.42 KW $Q =$ 29 VAR	$P =$ 1.43 KW $Q =$ 49 VAR	$P =$ 1.49 KW $Q =$ 25.5 VAR	$P =$ 1.43 KW $Q =$ 43 VAR
			Type-1 fuzzy	$P =$ 1.45 KW $Q =$ 100 VAR	$P =$ 1.46 KW $Q =$ 29 VAR	$P =$ 1.48 KW $Q =$ 32 VAR	$P =$ 1.45 KW $Q =$ 49 VAR	$P =$ 1.48 KW $Q =$ 48 VAR
			Type-2 fuzzy	$P =$ 1.45 KW $Q =$ 100 VAR	$P =$ 1.46 KW $Q =$ 30 VAR	$P =$ 1.48 KW $Q =$ 32 VAR	$P =$ 1.45 KW $Q =$ 50 VAR	$P =$ 1.49 KW $Q =$ 49 VAR

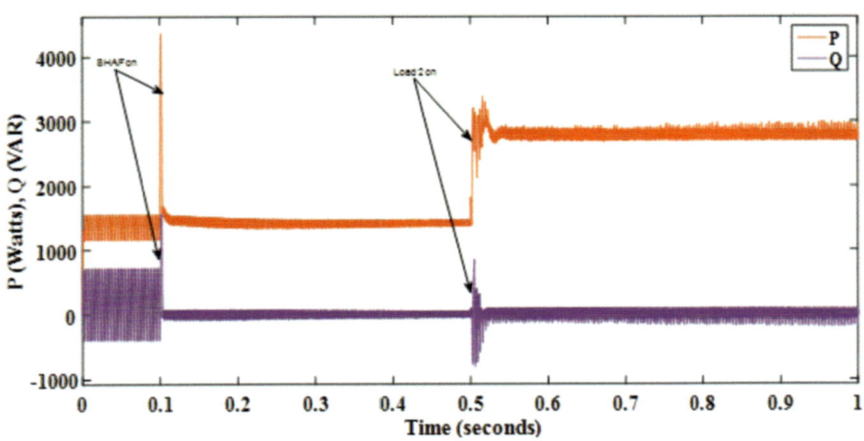

Figure 4.35 Real power and reactive power demands.

conditions. The waveforms in Figure 4.38 are concentrated on PI with PWM/HCC/AHCC/AFHCC type-1 technique respectively. Figure 4.38 (a) shows the load current without SHAF. It is observed that the load current is having power frequency as well as harmonics due to the non-linear

Table 4.15 Real and reactive powers of double-tuned SHAF with Fryze control under steady state.

Load	Reference controller	Without SAF	DC-link controller	With double-tuned SHAF (steady state)				
				PWM	HCC	AHCC	AFHCC type-1	AFHCC type-2
Diode bridge rectifier with RL load	Fryze	P = 1.5 KW Q = 720 VAR	PI	$P =$ 1.46 KW $Q =$ 60 VAR	$P =$ 1.42 KW $Q =$ 29 VAR	$P =$ 1.43 KW $Q =$ 49 VAR	$P =$ 1.49 KW $Q =$ 25.5 VAR	$P =$ 1.43 KW $Q =$ 43 VAR
			Type-1 fuzzy	$P =$ 1.45 KW $Q =$ 100 VAR	$P =$ 1.46 KW $Q =$ 29 VAR	$P =$ 1.48 KW $Q =$ 32 VAR	$P =$ 1.45 KW $Q =$ 50 VAR	$P =$ 1.48 KW $Q =$ 50 VAR
			Type-2 fuzzy	$P =$ 1.45 KW $Q =$ 100 VAR	$P =$ 1.46 KW $Q =$ 29 VAR	$P =$ 1.48 KW $Q =$ 32 VAR	$P =$ 1.45 KW $Q =$ 50 VAR	$P =$ 1.49 KW $Q =$ 50 VAR

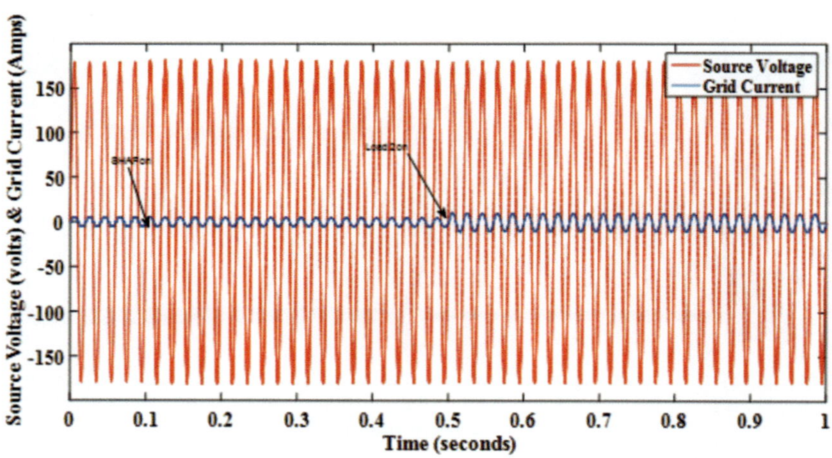

Figure 4.36 Power factor correction by the SHAF.

load (diode bridge rectifier load). The SHAF when switched on at 0.1 sec provides the desired compensating current to cancel out the harmonic current at PCC as shown in Figure 4.38 (b). By injecting equal but opposite

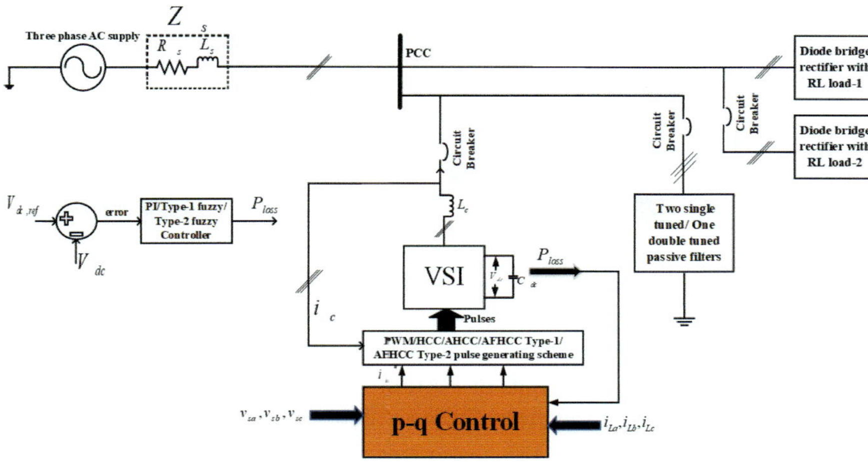

Figure 4.37 p-q power theory-based SHAF.

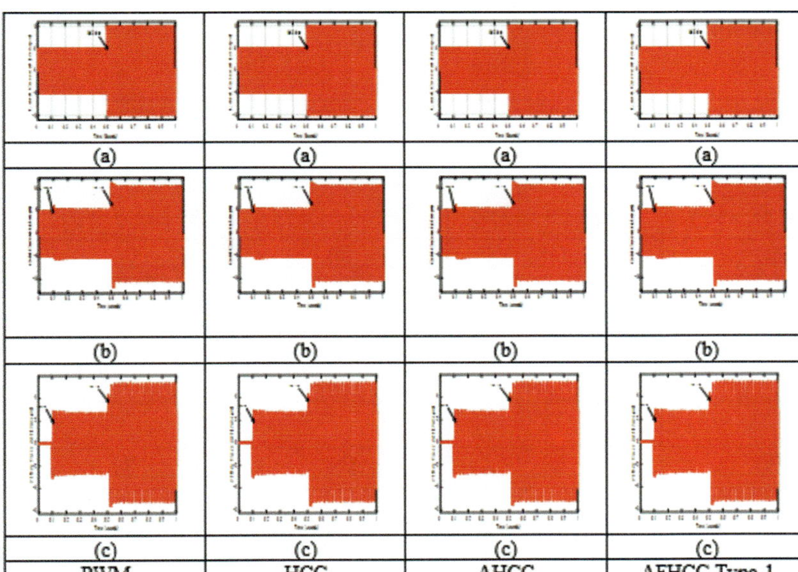

Figure 4.38 Simulation waveforms of (a) source current before compensation, (b) compensation currents, and (c) source current after compensation.

harmonic components, the harmonic compensation has been achieved. Figure 4.38 (c) shows the grid current after harmonic compensation. These waveforms are achieved under uncontrolled bridge rectifier load with steady-

and transient-state conditions for A-phase alone. The THD measurement of source/load current without filter under steady-state condition is 26.12% and under transient condition is 26.01%.

4.4.2 Case-2 – Type-1 fuzzy logic DC-link voltage controller

Type-1 fuzzy logic DC-link voltage controllers with PWM/HCC/AHCC/ AFHCC type-1/AFHCC type-2 are simulated under uncontrolled bridge rectifier with RL load. Figure 4.39 is achieved by using PWM/HCC/AHCC /AFHCCtype-1 under uncontrolled diode bridge rectifier load with steady state. Figure 4.39 (a) shows the source currents without compensation. Figure 4.39 (b) shows the compensating currents and Figure 4.39 (c) shows the grid currents after compensation. To mitigate the harmonic currents, the double-tuned SHAF (when it is ON) has to provide the desired compensating currents as shown in Figure 4.39 (b). The grid currents after compensation are shown in Figure 4.39 (c), the double-tuned SHAF is ON, the source currents become sinusoidal.

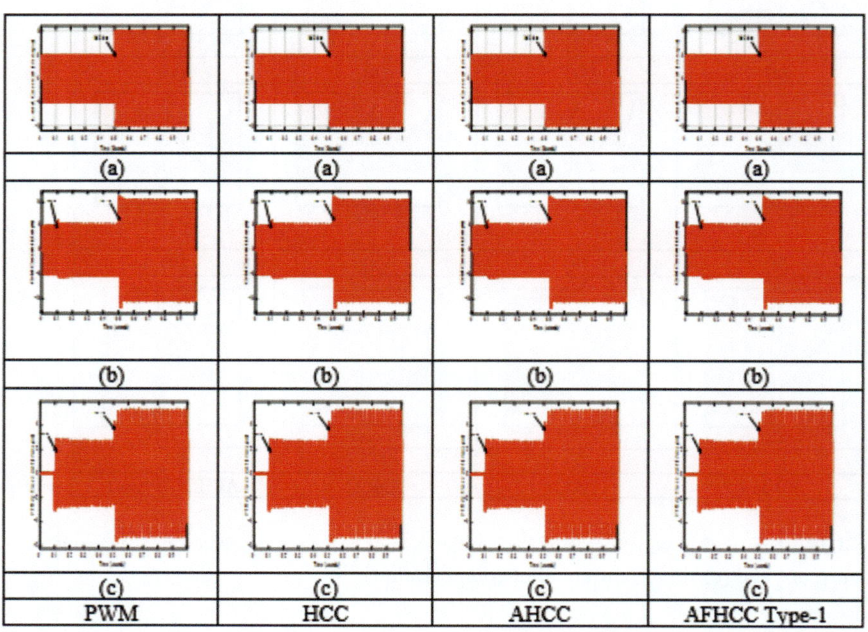

Figure 4.39 Simulation waveforms of (a) source current before compensation, (b) compensation currents, and (c) source current after compensation.

4.4.3 Case-3 – Type-2 fuzzy logic DC-link voltage controller

Type-2 fuzzy logic pulse generators with PWM/HCC/AHCC/AFHCC type-1/AFHCC type-2 are simulated under uncontrolled diode bridge rectifier with RL load. Figure 4.40 is achieved using PWM/HCC/AHCC/AFHCC type-1 under uncontrolled diode bridge rectifier load with steady- and transient-state conditions. Figure 4.40 (a) shows the grid currents without compensation. Figure 4.40 (b) represents the compensating currents and Figure 4.40 (c) shows the grid currents with SHAF compensation. To compensate the harmonic currents, the double-tuned SHAF has to give the desired compensating currents as represented in Figure 4.40 (b). After compensation, the grid currents are shown in Figure 4.40 (c), i.e., when the double-tuned SHAF is ON, the grid currents become mostly sinusoidal.

The grid current when SAF with type-2 fuzzy logic DC-link voltage controller and HCC switching scheme and double-tuned passive filter connected at 0.1 sec is as shown in Figure 4.41.

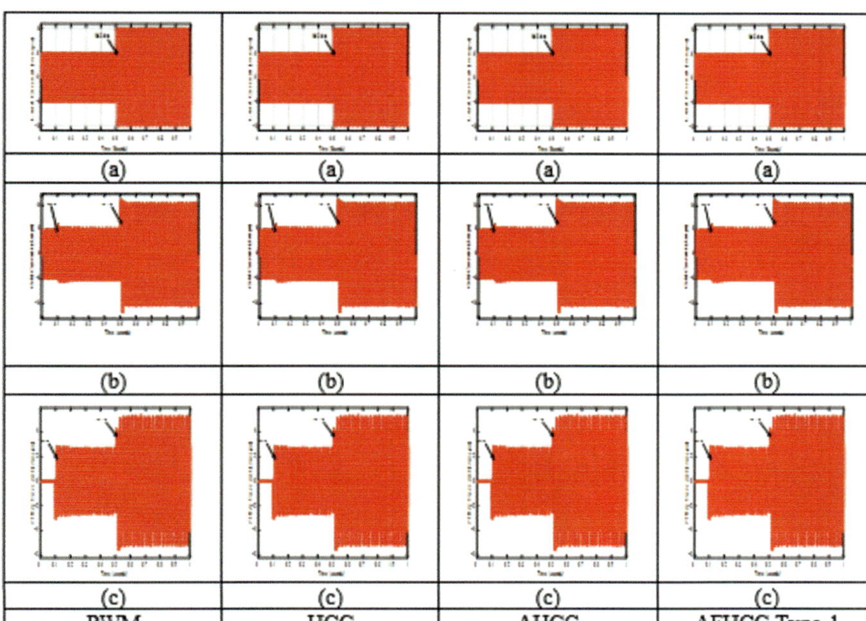

Figure 4.40 Simulation waveforms of (a) source current before compensation, (b) compensation currents, and (c) source current after compensation.

SAF compensation currents with type-2 fuzzy logic DC-link voltage controller and HCC switching scheme, and double tuned passive filter is as shown in Figure 4.42.

THD analysis: The harmonics against magnitude of the grid current is plotted with double-tuned SHAF under uncontrolled bridge rectifier with RL load. The simulation results are shown in Figure 4.43 is for type-2 fuzzy DC capacitor voltage controller and AFHCC type-2 pulse generator-based double-tuned SHAF under steady- and transient-state conditions with diode

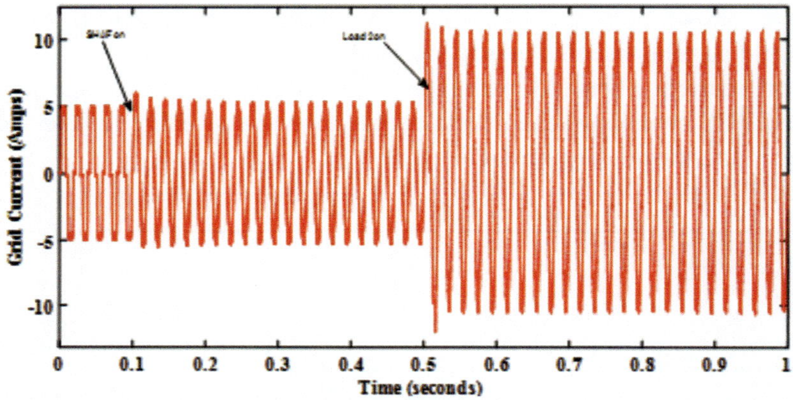

Figure 4.41 Double-tuned SHAF grid currents with PI DC-link voltage controller and HCC switching scheme.

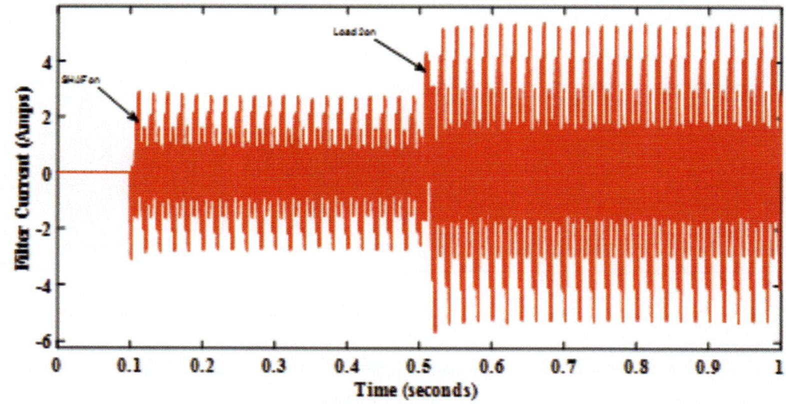

Figure 4.42 Double-tuned SHAF compensating currents with PI DC-link voltage controller and HCC switching scheme.

bridge rectifier. It has been observed that the grid current without SHAF is having harmonic components whereas with double-tuned SHAF, the THD fell from 26.12% to 1.06% in steady state and 26.21% to 1.16% in transient state respectively as represented in Figure 4.43.

Table 4.16 and Figure 4.44 show the THD measurement of source current with only SAF. The performance of SAF is analyzed with p-q control as a reference current controller, DC-link controller (PI, type-1 fuzzy, and type-2 fuzzy) and pulse generators (PWM, HCC, AHCC, AFHCC type-1, and AFHCC type-2). Similarly, Table 4.17 and Figure 4.45 show the THD measurement of source current when SAF with two single-tuned passive filters are connected to the system and Table 4.18 and Figure 4.46 show the THD measurement of source current when SAF with double-tuned passive filter is connected.

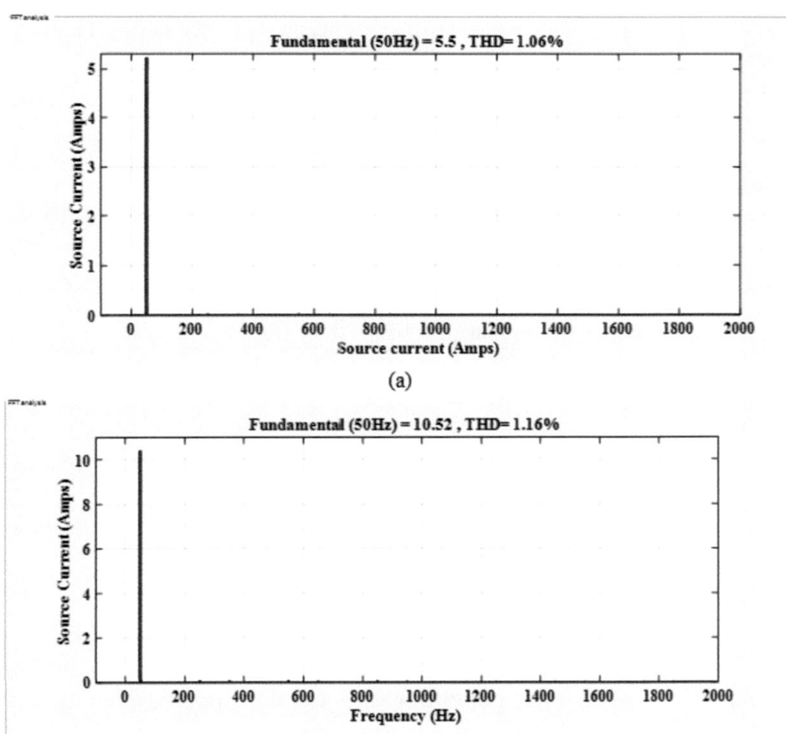

Figure 4.43 THD analysis of double-tuned SHAF (a) steady-state conditions (b) transient-state conditions.

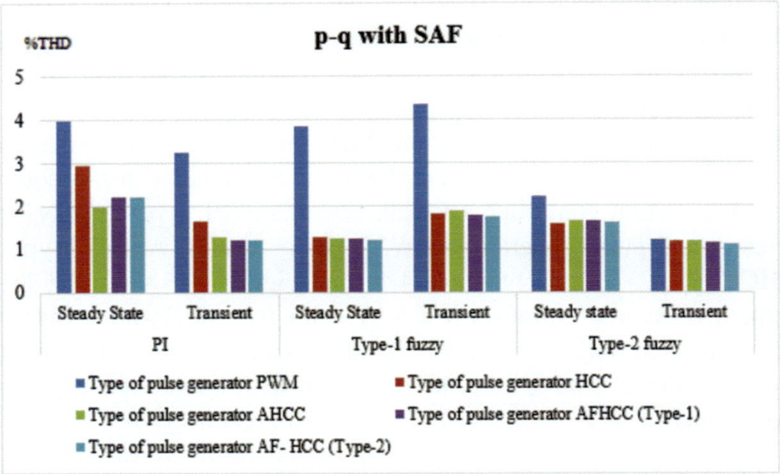

Figure 4.44 THD analysis of double-tuned SHAF (a) steady-state conditions (b) transient-state conditions.

Table 4.16 THD measurements of SAF with p-q control.

Load	Harmonic controller	DC-link controller	Condition	Type of pulse generator				
				PWM	HCC	AHCC	AF-HCC (type-1)	AF-HCC (type-2)
Diode bridge rectifier with RL load	*p-q*	PI	Steady State	3.99	2.94	1.99	2.22	2.22
			Transient	3.24	1.64	1.27	1.20	1.20
		Type-1 fuzzy	Steady State	3.85	1.28	1.25	1.23	1.22
			Transient	4.35	1.80	1.89	1.78	1.76
		Type-2 fuzzy	Steady state	2.20	1.57	1.64	1.63	1.62
			Transient	1.20	1.17	1.19	1.15	1.12

Figure 4.47 shows the DC-link voltage across the VSI capacitor bank of SHAF. In Figure 4.47, the SHAF utilizes type-2 fuzzy logic DC-link voltage controller, AFHCC type-2 pulse generator, and double-tuned passive filter. In a similar manner, all the DC-link voltage control techniques of SAF are able to maintain the constant DC-link voltage across VSI capacitor bank under steady- and transient-state conditions.

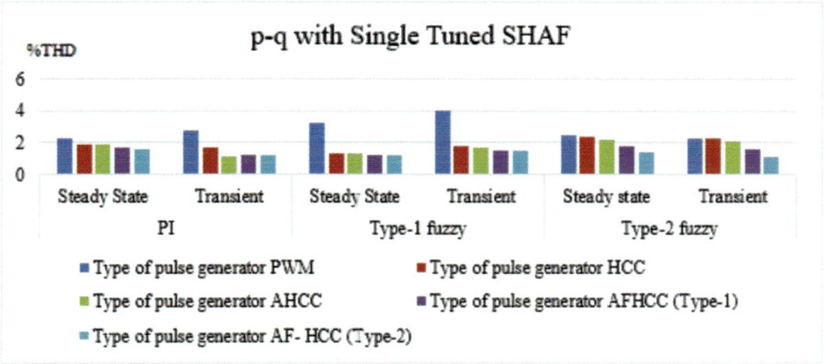

Figure 4.45 Comparison of single-tuned SHAF with various DC voltage controllers and pulse generators.

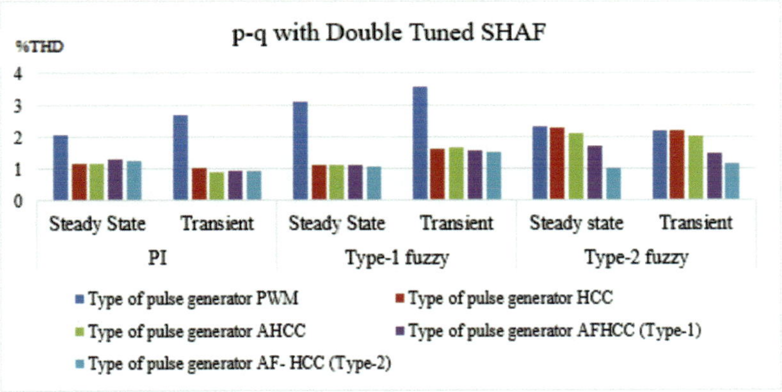

Figure 4.46 Comparison of single-tuned SHAF with various DC voltage controllers and pulse generators.

Table 4.17 THD measurements of single-tuned SHAF with p-q control.

Load	Harmonic controller	DC-link controller	Condition	Type of pulse generator				
				PWM	HCC	AHCC	AFHCC (type-1)	AF-HCC (type-2)
Diode bridge rectifier with RL load	*p-q*	PI	Steady State	2.26	1.87	1.87	1.68	1.64
			Transient	2.81	1.67	1.09	1.20	1.18
		Type-1 fuzzy	Steady State	3.24	1.31	1.32	1.25	1.24
			Transient	4	1.78	1.66	1.53	1.55
		Type-2 fuzzy	Steady state	2.46	2.38	2.2	1.8	1.44
			Transient	2.3	2.27	2.12	1.6	1.13

Table 4.18 THD measurements of double-tuned SHAF with p-q control.

Load	Harmonic con-troller	DC-link con-troller	Condition	Type of pulse generator				
				PWM	HCC	AHCC	AFHCC (type-1)	AF-HCC (type-2)
Diode bridge rectifier with RL load	*p-q*	PI	Steady State	2.07	1.19	1.19	1.30	1.28
			Transient	2.70	1.05	0.91	0.96	0.95
		Type-1 fuzzy	Steady State	3.12	1.13	1.14	1.12	1.09
			Transient	3.59	1.62	1.68	1.6	1.52
		Type-2 fuzzy	Steady state	2.35	2.3	2.15	1.7	**1.06**
			Transient	2.2	2.21	2.05	1.51	**1.16**

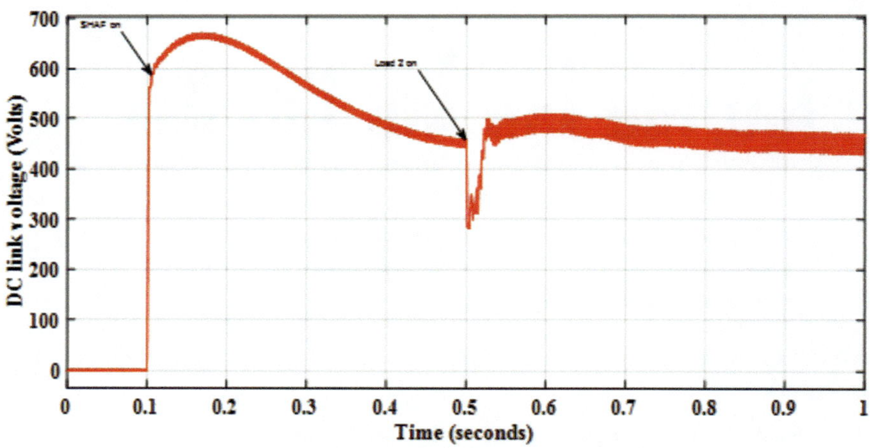

Figure 4.47 DC-link voltage across the VSI capacitor.

Figure 4.48 shows the reactive power balance, power factor correction, and real power demand under steady- and transient-state conditions using SHAF. In Figure 4.48, the SHAF utilizes type-2 fuzzy logic DC-link voltage controller, AFHCC type-2 pulse generator, and double-tuned passive filter. In a similar manner, the oscillating component of reactive power, real power, and power factor has been eliminated by all the proposed SAF, single-tuned and double-tuned SHAF.

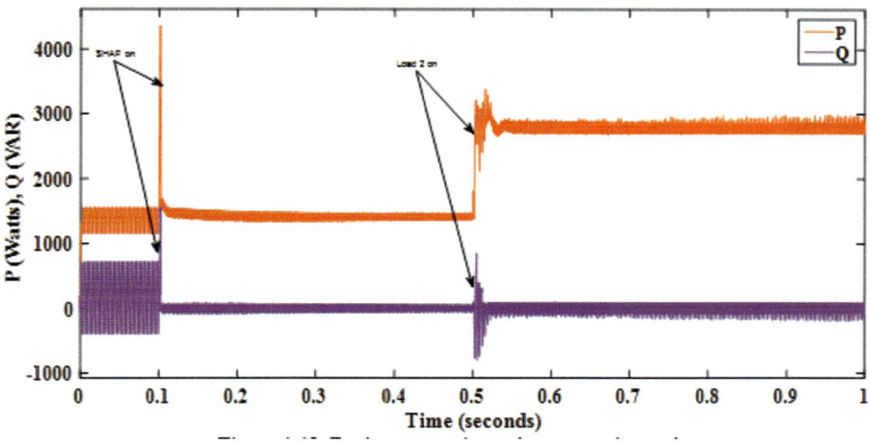

Figure 4.48 Real power and reactive power demands.

Real and Reactive power analysis: The double-tuned SHAF improves the power factor in distribution network. Figure 4.49 shows the waveform of power factor correction for A-phase voltage and current. The real and reactive powers are calculated as given in Tables 4.16–4.18. This result indicates that the SHAF is diminishing reactive power and therefore improving the power factor.

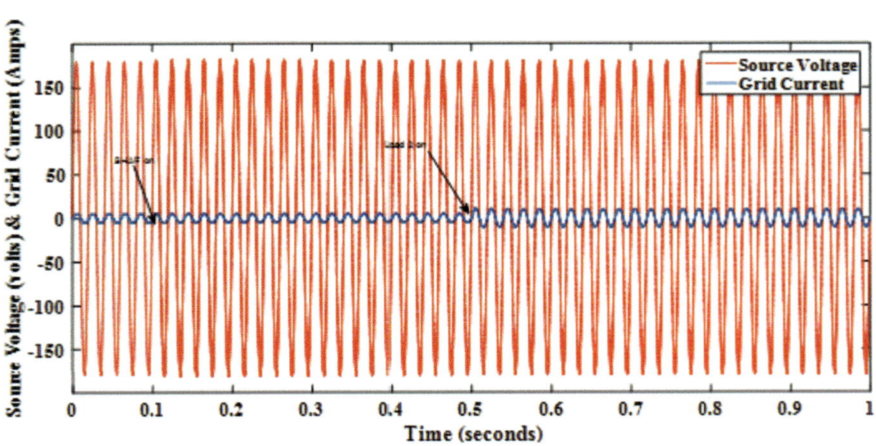

Figure 4.49 Power factor correction by the SHAF.

Table 4.19 Real and reactive powers of SAF with p-q control under steady state.

Load	Reference controller	Without SAF	DC-link controller	With SAF (steady state)				
				PWM	HCC	AHCC	AFHCC type-1	AFHCC type-2
Diode bridge rectifier with RL load	*p-q*	P = 1.5 KW Q = 720 VAR	PI	$P=$ 1.44 KW $Q=$ 60 VAR	$P=$ 1.51 KW $Q=29$ VAR	$P=$ 1.45 KW $Q=49$ VAR	$P=$ 1.49 KW $Q=$ 25.5 VAR	$P=$ 1.42 KW $Q=43$ VAR
			Type-1 fuzzy	$P=$ 1.45 KW $Q=$ 100 VAR	$P=$ 1.44 KW $Q=29$ VAR	$P=$ 1.49 KW $Q=31$ VAR	$P=$ 1.42 KW $Q=49$ VAR	$P=$ 1.49 KW $Q=49$ VAR
			Type-2 fuzzy	$P=$ 1.45 KW $Q=$ 100 VAR	$P=$ 1.46 KW $Q=29$ VAR	$P=$ 1.48 KW $Q=32$ VAR	$P=$ 1.55 KW $Q=49$ VAR	$P=$ 1.49 KW $Q=49$ VAR

Table 4.20 Real and reactive powers of single-tuned SHAF with p-q control under steady state.

Load	Reference controller	Without SAF	DC-link controller	With single-tuned SHAF (steady state)				
				PWM	HCC	AHCC	AFHCC type-1	AFHCC type-2
Diode bridge rectifier with RL load	*p-q*	P = 1.5 KW Q = 720 VAR	PI	$P=$ 1.45 KW $Q=$ 60 VAR	$P=$ 1.48 KW $Q=$ 29 VAR	$P=$ 1.43 KW $Q=$ 48 VAR	$P=$ 1.49 KW Q = 25.5 VAR	$P=$ 1.43 KW $Q=$ 43 VAR
			Type-1 fuzzy	$P=$ 1.45 KW $Q=$ 100 VAR	$P=$ 1.46 KW $Q=$ 29 VAR	$P=$ 1.48 KW $Q=$ 31 VAR	$P=$ 1.44 KW $Q=$ 50 VAR	$P=$ 1.49 KW $Q=$ 50 VAR
			Type-2 fuzzy	$P=$ 1.45 KW $Q=$ 100 VAR	$P=$ 1.45 KW $Q=$ 29 VAR	$P=$ 1.46 KW $Q=$ 32 VAR	$P=$ 1.46 KW $Q=$ 50 VAR	$P=$ 1.49 KW $Q=$ 50 VAR

Table 4.21 Real and reactive powers of double-tuned SHAF with p-q control under steady state.

Load	Reference controller	Without SAF	DC-link controller	With double-tuned SHAF (steady state)				
				PWM	**HCC**	**AHCC**	**AFHCC type-1**	**AFHCC type-2**
Diode bridge rectifier with RL load	*p-q*	P = 1.5 KW Q = 720 VAR	PI	$P=$ 1.46 KW $Q=$ 60 VAR	$P=$ 1.52 KW $Q=$ 29 VAR	$P=$ 1.45 KW $Q=$ 49 VAR	$P=$ 1.49 KW Q = 25.5 VAR	$P=$ 1.43 KW $Q=$ 43 VAR
			Type-1 fuzzy	$P=$ 1.45 KW $Q=$ 100 VAR	$P=$ 1.46 KW $Q=$ 29 VAR	$P=$ 1.48 KW $Q=$ 32 VAR	$P=$ 1.45 KW $Q=$ 50 VAR	$P=$ 1.48 KW $Q=$ 50 VAR
			Type-2 fuzzy	$P=$ 1.45 KW $Q=$ 100 VAR	$P=$ 1.46 KW $Q=$ 29 VAR	$P=$ 1.48 KW $Q=$ 32 VAR	$P=$ 1.45 KW $Q=$ 50 VAR	$P=$ 1.49 KW $Q=$ 50 VAR

Tables 4.19–4.21 give information regarding the real and reactive power requirements when connected to shunt active filter, single-tuned SHAF, and double-tuned SHAF to the non-linear system under steady-state conditions

Remarks: The p-q theory with various PWM pulse generators-based SHAF is simulated and compensation principle is evaluated. The AFHCC type-2 reduces the switching power losses and improves the SHAF performance than other techniques. From these results, it is proved that the THD of the grid current follows harmonic standards IEEE 519-2014.

4.5 Modified Real and Reactive Power (p-q) Theory

The modified p-q theory with different PWM pulse generators, DC capacitor voltage regulators, and tuned passive filters is performed by using MATLAB/Simulink tools. The model system is having three-phase VSI, coupling inductor, time-domain reference current extraction controller based on modified p-q theory, pulse generator, and tuned passive filters as represented in Figure 4.50. It is used to pull out the desired reference current

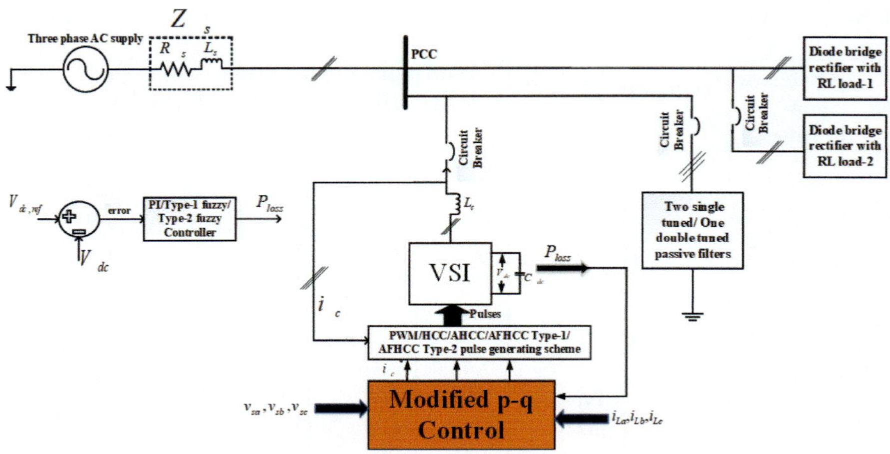

Figure 4.50 Modified p-q power theory-based SHAF.

from the polluted load current. The reference current is related with original compensation current to produce the trigger pulses for SAF using PWM/HCC/AHCC/AFHCC type-1/AFHCC type-2 controllers. The SAF generates the anti-harmonic current to cancel out current harmonics and balance reactive power in the distributed network. The whole system is tested under steady- and transient-state conditions. Further, the tuned passive filters like single-tuned and double-tuned combine with the SAF that gives better performance.

4.5.1 Case-1 – PI DC-link voltage controller

PI DC-link voltage controller with different pulse generators is simulated for SHAF under balanced uncontrolled diode bridge rectifiers with RL load conditions. The waveforms in Figure 4.51 are concentrated on PI with PWM/HCC/AHCC/AFHCC type-1 technique respectively. Figure 4.51 (a) shows the load current before without SHAF. It is observed that the load current is having power frequency as well as harmonics due to the non-linear load (diode bridge rectifier load). The PWM/HCC/AHCC/AFHCC type-1/AFHCC type-2 pulse-generating techniques have been used to drive the SAF. The SHAF when switched on at 0.1 sec, provides the desired compensating current to cancel out the harmonic current at PCC as shown in Figure 4.51 (b). By injecting equal but opposite harmonic components, the harmonic compensation has been achieved. Figure 4.51 (c) shows the

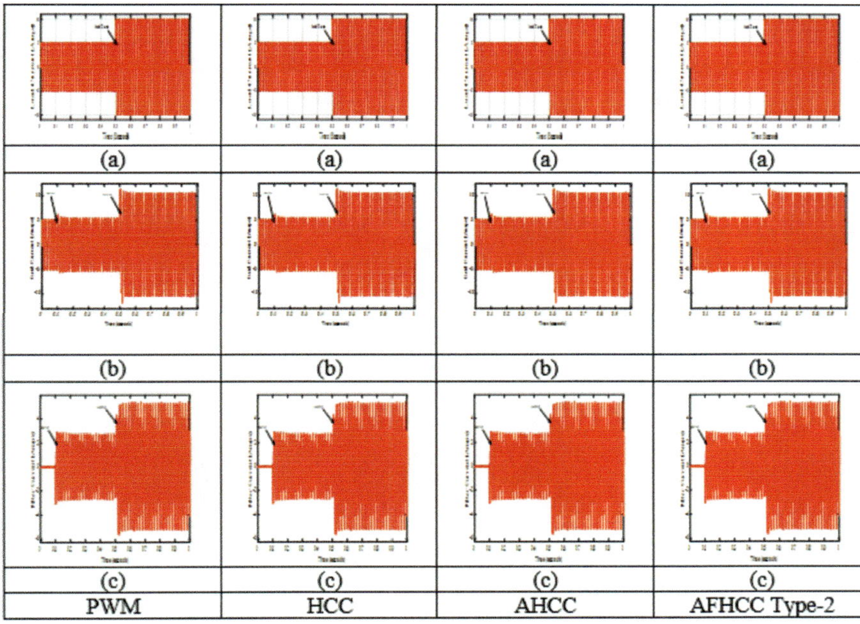

Figure 4.51 Simulation waveforms of (a) source current before compensation, (b) compensation currents, and (c) source current after compensation.

grid current after harmonic compensation. These waveforms are achieved under uncontrolled bridge rectifier load with steady- and transient-state conditions and these are for A-phase alone; B- and C-phase are phase shifted by 120°. The THD measurement of source/load current without filter under steady-state condition is 26.12% and under transient condition is 26.01%.

4.5.2 Case-2 – Type-1 fuzzy logic DC-link voltage controller

Type-1 fuzzy logic DC-link voltage controllers with PWM/HCC/AHCC /AFHCC type-1/AFHCC type-2 are simulated under uncontrolled bridge rectifier with RL load. Figure 4.52 is achieved by using PWM/HCC/AHCC /AFHCC type-1 under uncontrolled diode bridge rectifier load with steady state. Figure 4.52 (a) shows the source currents without compensation. Figure 4.52 (b) shows the compensating currents and Figure 4.52 (c) shows the grid currents after compensation. To mitigate the harmonic currents, the double-tuned SHAF (when it is ON) has to provide the desired compensating currents as shown in Figure 4.52 (b). The grid currents after compensation are

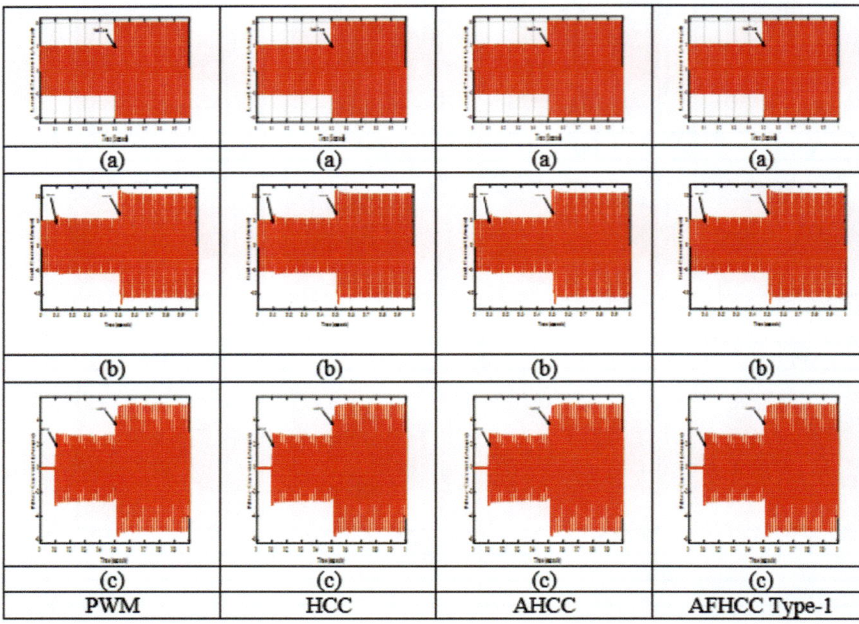

Figure 4.52 Simulation waveforms of (a) source current before compensation, (b) compensation currents, and (c) source current after compensation.

shown in Figure 4.52 (c), the double-tuned SHAF is ON, the source currents become sinusoidal.

4.5.3 Case-3 – Type-2 fuzzy logic DC-link voltage controller

Type-2 fuzzy logic DC-link voltage controllers with PWM/HCC/AHCC /AFHCC type-1/AFHCC type-2 are simulated under uncontrolled diode bridge rectifier with RL load. Figure 4.53 is achieved using PWM/HCC/ AHCC/AFHCC type-1 under uncontrolled diode bridge rectifier load with steady state. Figure 4.53 (a) shows the grid currents without compensation. Figure 4.53 (b) represents the compensating currents and Figure 4.53 (c) shows the grid currents with SHAF compensation. To compensate the harmonic currents, the double-tuned SHAF has to give the desired compensating currents as represented in Figure 4.53 (b). After compensation, the grid currents are shown in Figure 4.53 (c), i.e., when the double-tuned SHAF is ON, the grid currents become mostly sinusoidal.

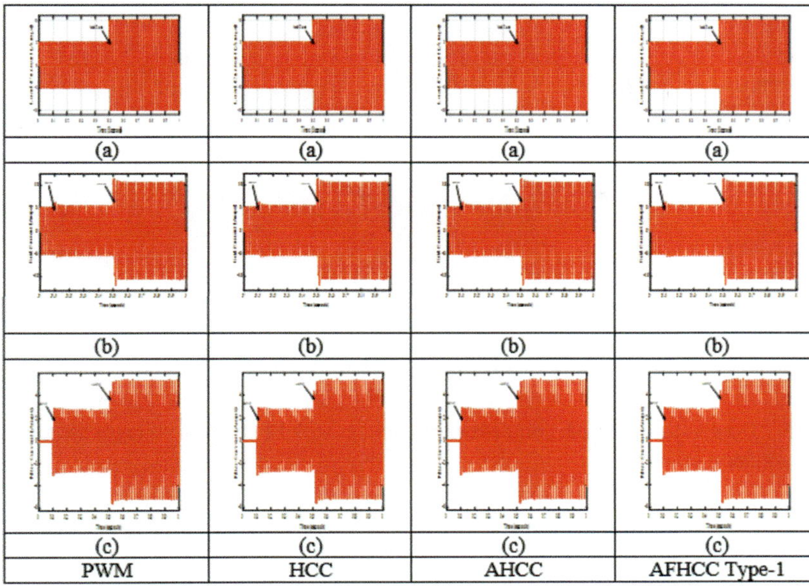

Figure 4.53 Simulation waveforms of (a) source current before compensation, (b) compensation currents, and (c) source current after compensation.

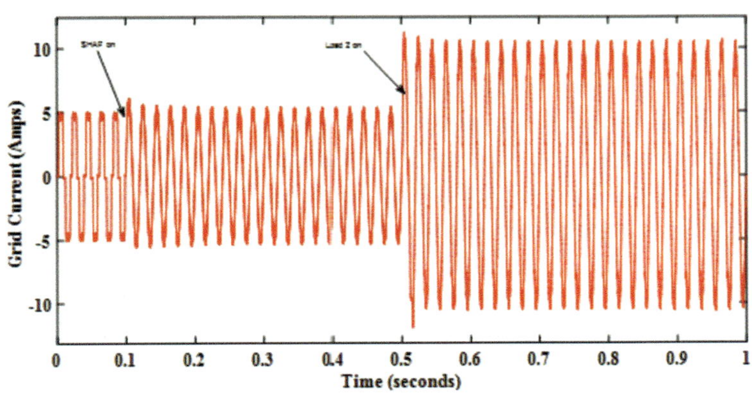

Figure 4.54 Double-tuned SHAF grid currents with PI DC-link voltage controller and HCC switching scheme.

Figure 4.54 shows the grid current of phase A, when SAF with type-2 fuzzy logic DC-link voltage controller and AHCC switching scheme is connected to double-tuned passive filter at 0.1 sec.

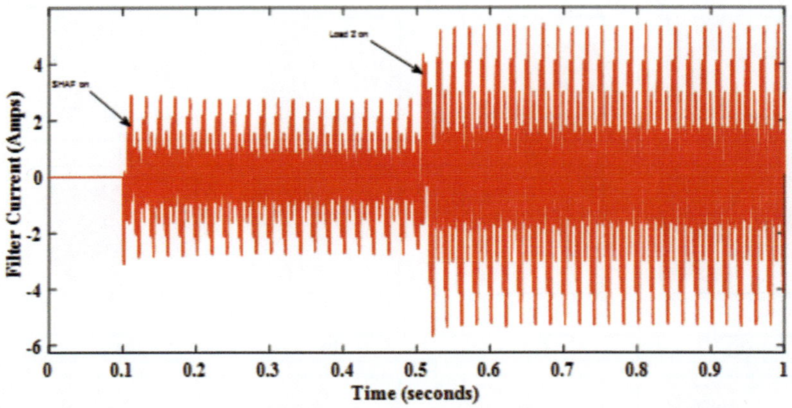

Figure 4.55 Double-tuned SHAF compensating currents with PI DC-link voltage controller and HCC switching scheme.

The SAF compensation currents with type-2 fuzzy logic DC-link voltage controller, AHCC switching scheme and double-tuned passive filter is as shown in Figure 4.55.

THD analysis: The harmonics against magnitude of the grid current is plotted with double-tuned SHAF under uncontrolled bridge rectifier with RL load. The simulation results shown in Figure 4.56 are for type-2 fuzzy DC capacitor voltage controller and AFHCC type-2 pulse generator-based double-tuned SHAF under steady- and transient-state conditions with diode bridge rectifier. It has been observed that the grid current without SHAF is having harmonic components whereas with double-tuned SHAF the THD fell from 26.12% to 0.91% in steady state and from 26.21% to 0.81% in transient state respectively as represented in Figure 4.56.

Table 4.22 THD measurements of SAF with modified p-q control.

Load	Harmonic controller	DC-link controller	Condition	Type of pulse generator				
				PWM	HCC	AHCC	AFHCC (type-1)	AF-HCC (type-2)
Diode bridge rectifier with RL load	Modified *p-q*	PI	Steady State	3.45	2.08	2.05	2.08	2.05
			Transient	2.85	1.18	1.20	1.36	1.35
		Type-1 fuzzy	Steady State	3.87	1.12	1.9	1.96	1.35
			Transient	4.26	1.1	1.85	1.82	1.28
		Type-2 fuzzy	Steady state	3.6	1.08	1.82	1.6	1.21
			Transient	3.55	1.1	1.8	1.56	1.28

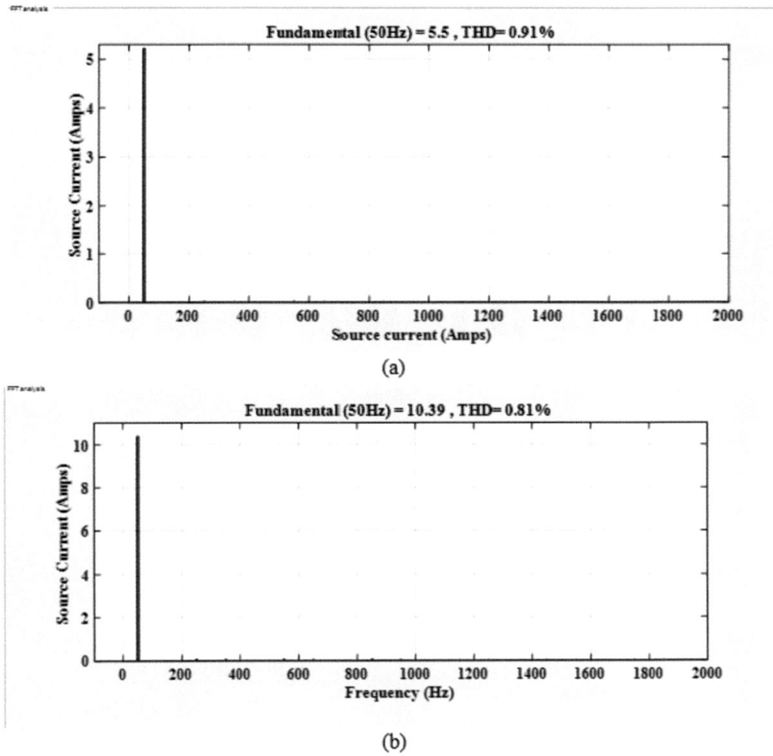

Figure 4.56 THD analysis of double-tuned SHAF (a) steady-state conditions (b) transient-state conditions.

Table 4.23 THD measurements of single-tuned SHAF with modified p-q control.

Load	**Harmonic controller**	**DC-link controller**	**Condition**	**Type of pulse generator**				
				PWM	**HCC**	**AHCC**	**AFHCC (type-1)**	**AF-HCC (type-2)**
Diode bridge rectifier with RL load	Modified *p-q*	PI	Steacy State	3.11	1.87	1.86	1.68	1.64
			Transient	3.24	1.09	1.10	1.23	1.2
		Type-1 fuzzy	Steacy State	3.2	1.8	1.78	1.62	1.5
			Transient	3.24	1.75	1.7	1.6	1.32
		Type-2 fuzzy	Steady state	3.18	1.7	1.68	1.56	0.92
			Transient	3.2	1.69	1.6	1.5	0.87

Table 4.22 and Figure 4.57 show the THD measurement of source current with only SAF. The performance of SAF is analyzed with modified p-q power

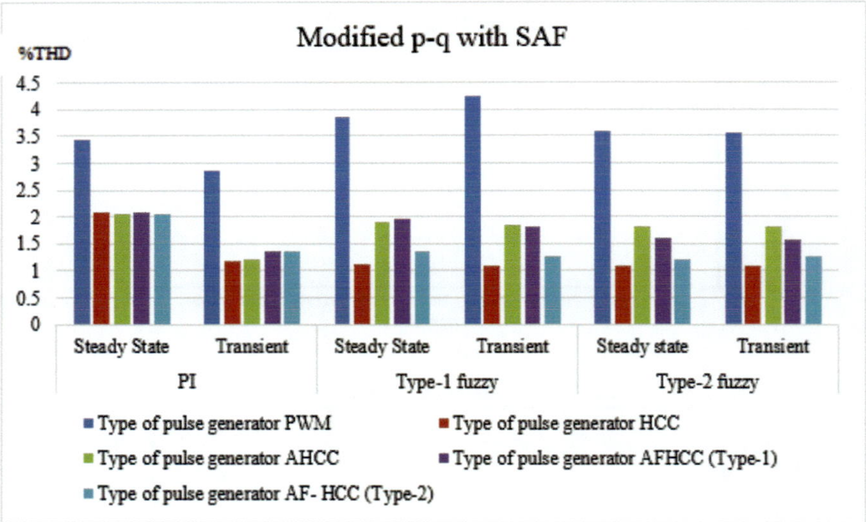

Figure 4.57 Comparison of SAF with various DC voltage controllers and pulse generators.

Table 4.24 THD measurements of double-tuned SHAF with modified p-q control.

Load	**Harmonic controller**	**DC-link controller**	**Condition**	**Type of pulse generator**				
				PWM	**HCC**	**AHCC**	**AFHCC (type-1)**	**AF-HCC (type-2)**
Diode bridge rectifier with RL load	Modified *p-q*	PI	Steady State	3.01	1.19	1.2	1.23	1.21
			Transient	4.61	1.15	1.12	1.1	0.97
		Type-1 fuzzy	Steady State	2.68	1.12	1.1	1.12	0.96
			Transient	2.72	1.1	1.05	1.08	0.95
		Type-2 fuzzy	Steady state	2.58	1.05	1	1.05	**0.91**
			Transient	2.52	1	0.99	0.98	**0.81**

control as a reference current controller, DC-link controller (PI, type-1 fuzzy, and type-2 fuzzy), and pulse generators (PWM, HCC, AHCC, AFHCC type-1 and AFHCC type-2). Similarly, Table 4.23 and Figure 4.58 show the THD measurement of source current when SAF with two single-tuned passive filters is connected to the system and Table 4.24 and Figure 4.59 show the THD measurement of source current when SAF with double-tuned passive filter is connected.

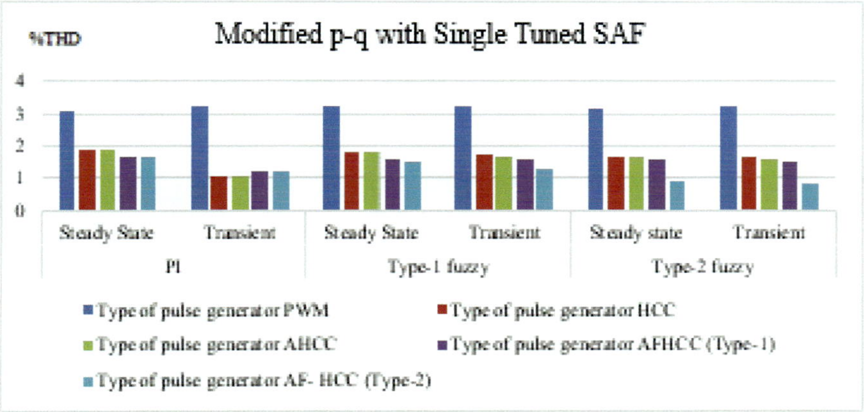

Figure 4.58 Comparison of single-tuned SHAF with various DC voltage controllers and pulse generators.

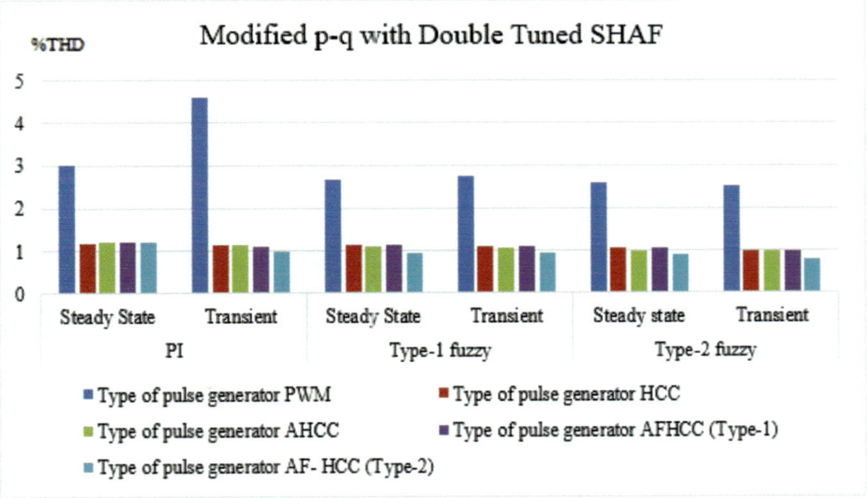

Figure 4.59 Comparison of double-tuned SHAF with various DC voltage controllers and pulse generators.

Real and Reactive power analysis: The double-tuned SHAF improves the power factor in distribution network. Figure 4.60 shows the waveform of power factor correction for A-phase voltage and current. The real and reactive powers are calculated as given in Tables 4.22 to 4.24. This result indicates that, the SHAF is diminishing reactive power and therefore improving the power factor.

Tables 4.25 to 4.27 give the information regarding the real and reactive power requirements when connected to shunt active filter, single-tuned SHAF, and double-tuned SHAF to the non-linear system under transient-state conditions.

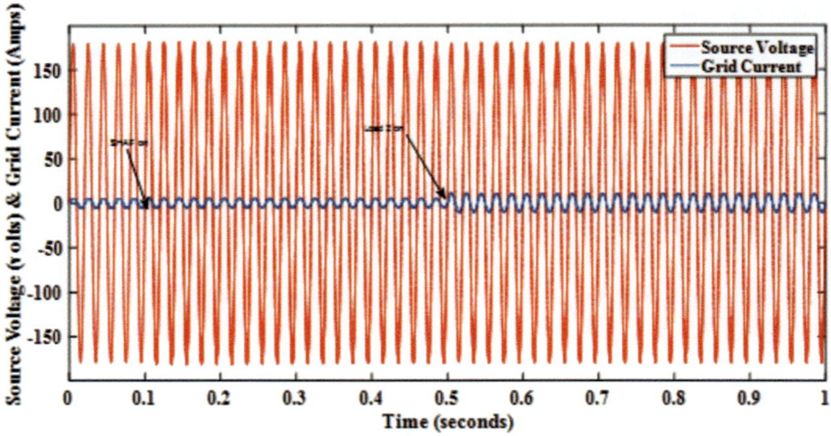

Figure 4.60 Power factor correction by the SHAF.

Table 4.25 Real and reactive powers of SAF with modified *p-q* control under transient state.

Load	Reference controller	Without SAF	DC-link controller	With SAF (steady state)				
				PWM	**HCC**	**AHCC**	**AFHCC type-1**	**AFHCC type-2**
Diode bridge rectifier with RL load	Modified *p-q*	$P = 1.5$ KW $Q = 1540$ VAR	PI	$P =$ 2.8 KW $Q =$ 90 VAR	$P =$ 2.81 KW $Q =$ 92 VAR	$P =$ 2.78 KW $Q =$ 91 VAR	$P =$ 2.75 KW $Q = 92$ VAR	$P =$ 2.79 KW $Q =$ 91 VAR
			Type-1 fuzzy	$P =$ 2.8 KW $Q =$ 90 VAR	$P =$ 2.81 KW $Q =$ 92 VAR	$P =$ 2.78 KW $Q =$ 91 VAR	$P =$ 2.85 KW $Q =$ 92 VAR	$P =$ 2.79 KW $Q =$ 91 VAR
			Type-2 fuzzy	$P =$ 2.89 KW $Q =$ 90 VAR	$P =$ 2.81 KW $Q =$ 92 VAR	$P =$ 2.78 KW $Q =$ 91 VAR	$P =$ 2.75 KW $Q =$ 92 VAR	$P =$ 2.79 KW $Q =$ 91 VAR

Figure 4.61 shows the DC-link voltage across the VSI capacitor bank of SHAF. In Figure 4.61, the SHAF utilizes type-2 fuzzy logic DC-link voltage

Table 4.26 Real and reactive powers of single-tuned SHAF with modified p-q under transient state.

Load	**Reference controller**	**Without SAF**	**DC-link controller**	**With single-tuned SHAF (transient state)**				
				PWM	**HCC**	**AHCC**	**AFHCC type-1**	**AFHCC type-2**
Diode bridge rectifier with RL load	Modified *p-q*	P = 1.5 KW Q = 1540 VAR	PI	$P=$ 2.8 KW $Q=$ 904 VAR	$P=$ 2.81 KW $Q=$ 92 VAR	$P=$ 2.78 KW $Q=$ 91 VAR	$P=$ 2.75 KW $Q=$ 91 VAR	$P=$ 2.79 KW $Q=$ 90 VAR
			Type-1 fuzzy	$P=$ 2.8 KW $Q=$ 90 VAR	$P=$ 2.81 KW $Q=$ 92 VAR	$P=$ 2.78 KW $Q=$ 91 VAR	$P=$ 2.75 KW $Q=$ 92 VAR	$P=$ 2.79 KW $Q=$ 91 VAR
			Type-2 fuzzy	$P=$ 2.8 KW $Q=$ 90 VAR	$P=$ 2.81 KW $Q=$ 92 VAR	$P=$ 2.78 KW $Q=$ 91 VAR	$P=$ 2.75 KW $Q=$ 92 VAR	$P=$ 2.79 KW $Q=$ 91 VAR

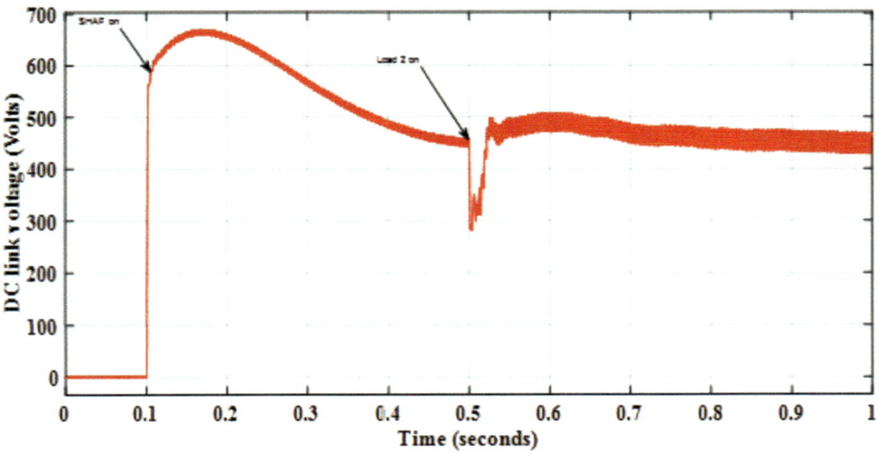

Figure 4.61 DC voltage across the VSI capacitor link.

Table 4.27 Real and reactive powers of double-tuned SHAF with modified p-q under transient state.

Load	Reference controller	Without SAF	DC-link controller	With double-tuned SHAF (transient state)				
				PWM	HCC	AHCC	AFHCC type-1	AFHCC type-2
Diode bridge rectifier with RL load	Modified *p-q*	P = 1.5 KW Q = 1540 VAR	PI	$P =$ 2.8 KW $Q =$ 90 VAR	$P =$ 2.81 KW $Q =$ 92 VAR	$P =$ 2.78 KW $Q =$ 91 VAR	$P =$ 2.75 KW $Q =$ 92 VAR	$P =$ 2.79 KW $Q =$ 91 VAR
			Type-1 fuzzy	$P =$ 2.8 KW $Q =$ 90 VAR	$P =$ 2.81 KW $Q =$ 92 VAR	$P =$ 2.78 KW $Q =$ 91 VAR	$P =$ 2.75 KW $Q =$ 92 VAR	$P =$ 2.79 KW $Q =$ 91 VAR
			Type-2 fuzzy	$P =$ 2.7 KW $Q =$ 90 VAR	$P =$ 2.81 KW $Q =$ 92 VAR	$P =$ 2.75 KW $Q =$ 91 VAR	$P =$ 2.75 KW $Q =$ 92 VAR	$P =$ 2.79 KW $Q =$ 91 VAR

controller, AFHCC type-2 pulse generator, and double-tuned passive filter. In a similar manner, all the DC-link control techniques of SAF are able to maintain the constant DC-link voltage across VSI capacitor bank under steady- and transient-state conditions.

Figure 4.62 shows the reactive power balance, power factor correction, and real power demand under steady- and transient-state condition using SHAF. In Figure 4.62, the SHAF utilizes type-2 fuzzy logic DC-link voltage controller, AFHCC type-2 pulse generator, and double-tuned passive filter. In a similar manner, the oscillating component of reactive power, real power, and power factor has been eliminated by all the proposed SAF, single-tuned and double-tuned SHAF.

Remarks: The modified p-q theory with various PWM pulse generators-based SHAF is simulated and compensation principle is evaluated. The AFHCC type-2 reduces the switching power losses and improves the SHAF performance than other techniques. From these results, it is proved that the THD of the grid current is following harmonic standards IEEE 519-2014.

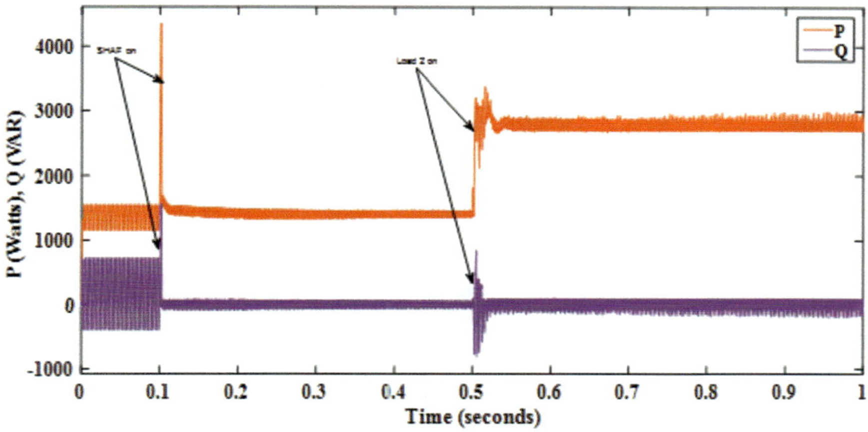

Figure 4.62 Real power and reactive power demands.

4.6 Synchronous Reference Frame (SRF) theory

The SRF theory with different PWM pulse generators, DC capacitor voltage regulators, and tuned passive filters is performed by using MATLAB/Simulink tools. The model system is having three-phase VSI, coupling inductor, time-domain reference current extraction controller based on SRF theory, pulse generator, and tuned passive filters as represented in Figure 4.63. The reference current is related with original compensation current to produce the trigger pulses for SAF using PWM/HCC/AHCC/AFHCC type-1/AFHCC type-2 controllers. The SAF generates the anti-harmonic current to cancel out current harmonics and balance reactive power in the distributed network. The whole system is tested under steady- and transient-state conditions. Further, the tuned passive filters like single- and double-tuned combine with the SAF that gives better performance.

4.6.1 Case-1 – PI DC-link voltage controller

PI DC-link voltage controller with different pulse generators is simulated for SHAF under balanced uncontrolled diode bridge rectifiers with RL-load conditions. The waveforms in Figure 4.64 are concentrated on PI with PWM/HCC/AHCC/AFHCC type-2 techniques respectively. Figure 4.64 (a) shows the load current without SHAF. It is observed that the load current is having power frequency as well as harmonics due to the nonlinear load (diode bridge rectifier load). The PWM/HCC/AHCC/AFHCC

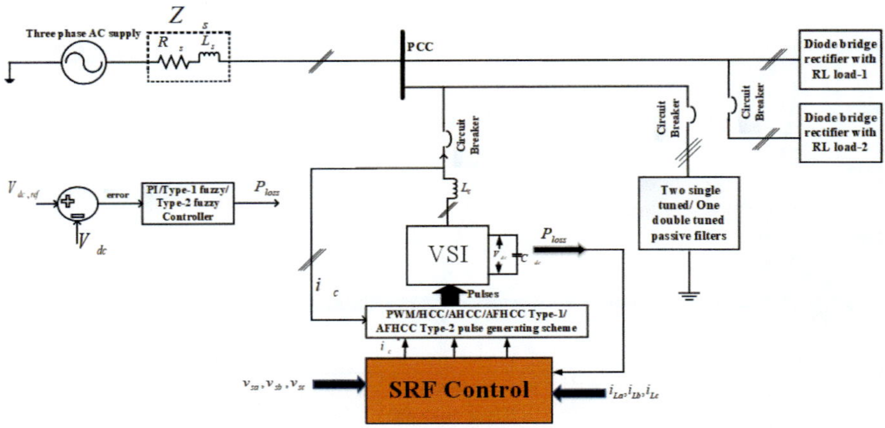

Figure 4.63 SRF theory-based SHAF.

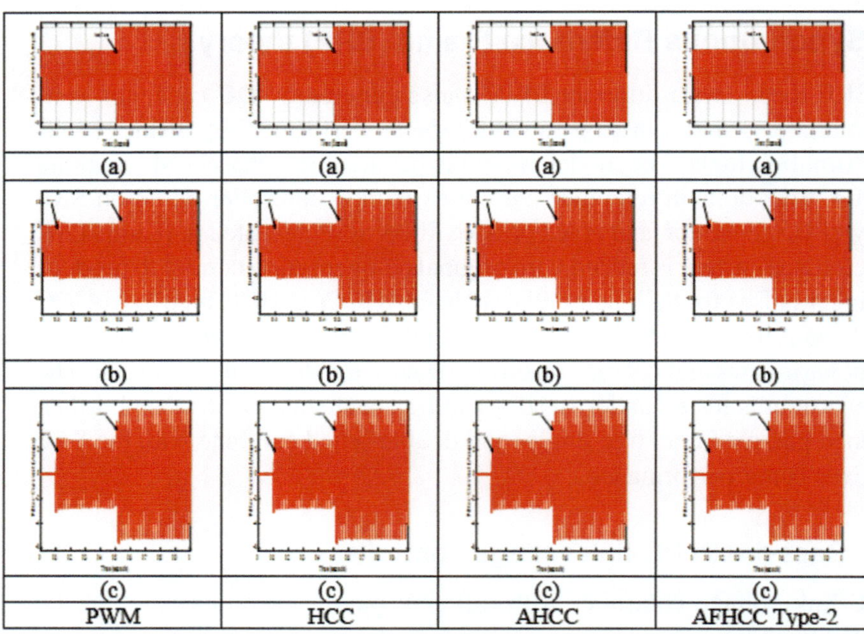

Figure 4.64 Simulation waveforms of (a) source current before compensation, (b) compensation currents, and (c) source current after compensation.

type-1/AFHCC type-2 pulse-generating techniques have been used to drive the SAF. The SHAF when switched ON at 0.1 sec provides the desired compensating current to cancel out the harmonic current at PCC as shown in Figure 4.64 (b). By injecting equal but opposite harmonic components, the harmonic compensation has been achieved. Figure 4.64 (c) shows the grid current after harmonic compensation. These waveforms are achieved under uncontrolled bridge rectifier load with steady and transient conditions and these are for A-phase alone. The THD measurement of source/load current without filter under steady-state condition is 26.12% and under transient-state condition is 26.01%.

4.6.2 Case-2 – Type-1 fuzzy logic DC-link voltage controller

Type-1 fuzzy logic DC-link voltage controllers with PWM/HCC/AHCC /AFHCCtype-1/AFHCC type-2 are simulated under uncontrolled bridge rectifier with RL load. Figure 4.65 is achieved by using PWM/HCC/AHCC /AFHCC type-2 under uncontrolled diode bridge rectifier load with steady and transient state. Figure 4.65 (a) shows the source currents without compensation. Figure 4.65 (b) shows the compensating currents and Figure 4.65 (c) shows the grid currents after compensation. To mitigate the

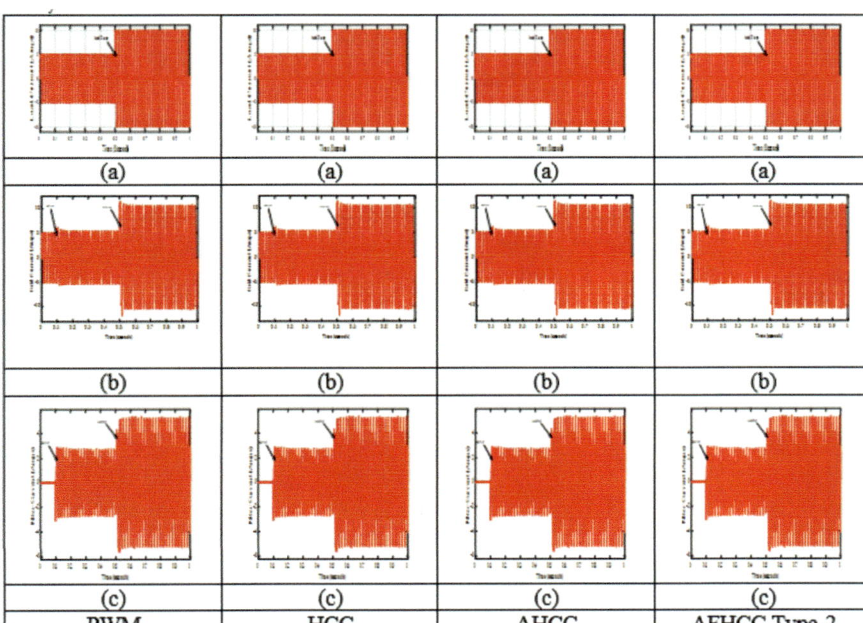

Figure 4.65 Simulation waveforms of (a) source current before compensation, (b) compensation currents, and (c) source current after compensation.

harmonic currents, the double-tuned SHAF (when it is ON) has to provide the desired compensating currents as shown in Figure 4.65 (b). The grid currents after compensation are shown in Figure 4.65 (c), the double-tuned SHAF is ON, the source currents become sinusoidal.

Figure 4.66 shows the grid current of phase A, when SAF with type-2 fuzzy logic DC-link voltage controller and AHCC switching scheme connected to double-tuned passive filter at 0.1 sec.

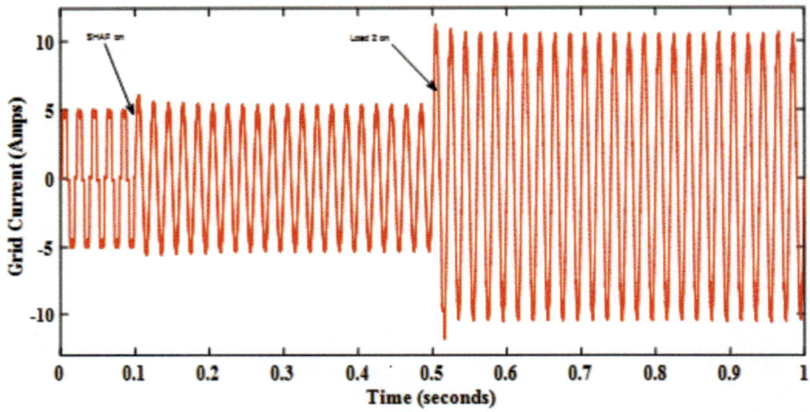

Figure 4.66 Double-tuned SHAF grid currents with PI DC-link voltage controller and HCC switching scheme.

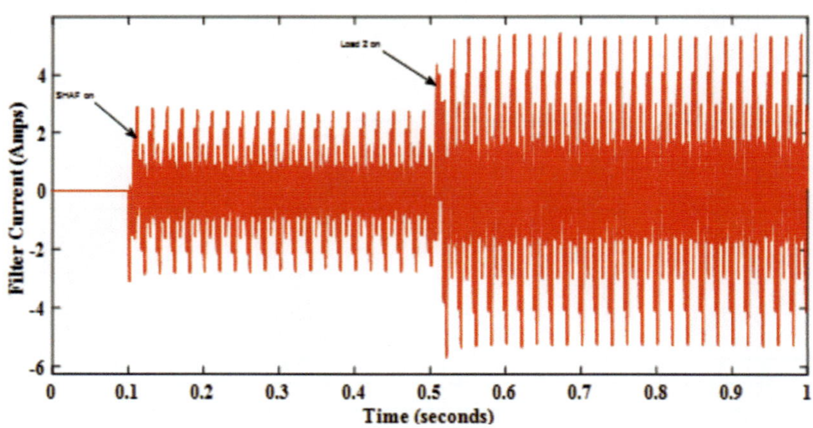

Figure 4.67 Double-tuned SHAF grid currents with PI DC-link voltage.

Figure 4.67 shows the SAF compensation currents with type-2 fuzzy logic DC-link voltage controller and AHCC switching scheme and double-tuned passive filter.

4.6.3 Case-3 – Type-2 fuzzy logic DC-link voltage controller

Type-2 fuzzy logic DC-link voltage controllers with PWM/HCC/AHCC /AFHCC type-2/AFHCC type-2 are simulated under uncontrolled diode bridge rectifier with RL load. Figure 4.68 is achieved using PWM/HCC /AHCC/AFHCC type-2 under uncontrolled diode bridge rectifier load with steady and transient state. Figure 4.68 (a) shows the grid currents without compensation. Figure 4.68 (b) represents the compensating currents and Figure 4.68 (c) shows the grid currents with SHAF compensation. To compensate the harmonic currents, the double-tuned SHAF has to give the desired compensating currents as represented in Figure 4.68 (b). After compensation, the grid currents are shown in Figure 4.68 (c), i.e., when the double-tuned SHAF is ON, the grid currents become mostly sinusoidal.

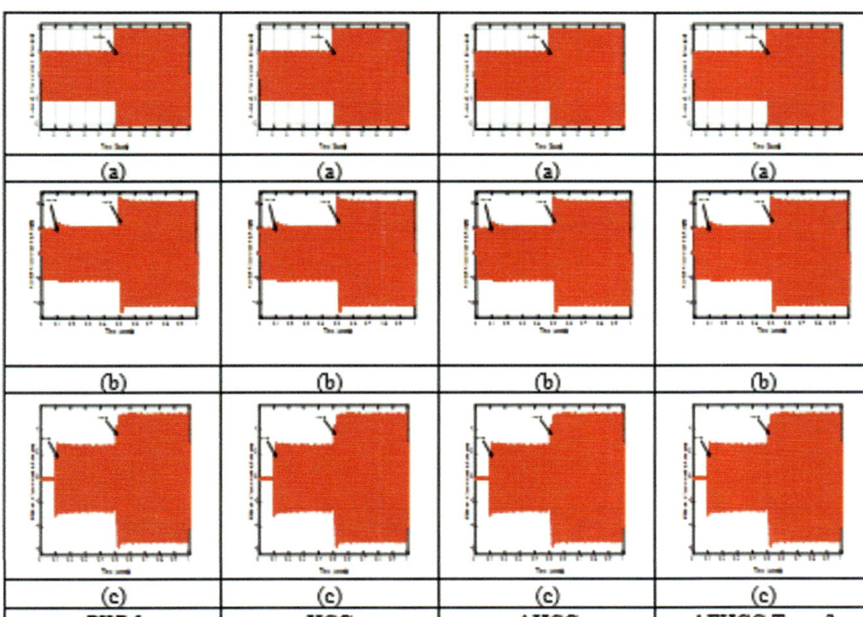

Figure 4.68 Simulation waveforms of (a) source current before compensation, (b) compensation currents, and (c) source current after compensation.

THD analysis: The harmonics against magnitude of the grid current is plotted with double-tuned SHAF under uncontrolled bridge rectifier with RL load. The simulation results shown in Figure 4.69 are for type-2 fuzzy DC capacitor voltage controller and AFHCC type-2 pulse generator-based double-tuned SHAF under steady- and transient-state conditions with diode bridge rectifier. It has been observed that the grid current without SHAF is having harmonic components whereas with double-tuned SHAF the THD fell from 26.12% to 2% in steady state and from 26.21% to 2.03% in transient-state respectively as represented in Figure 4.69.

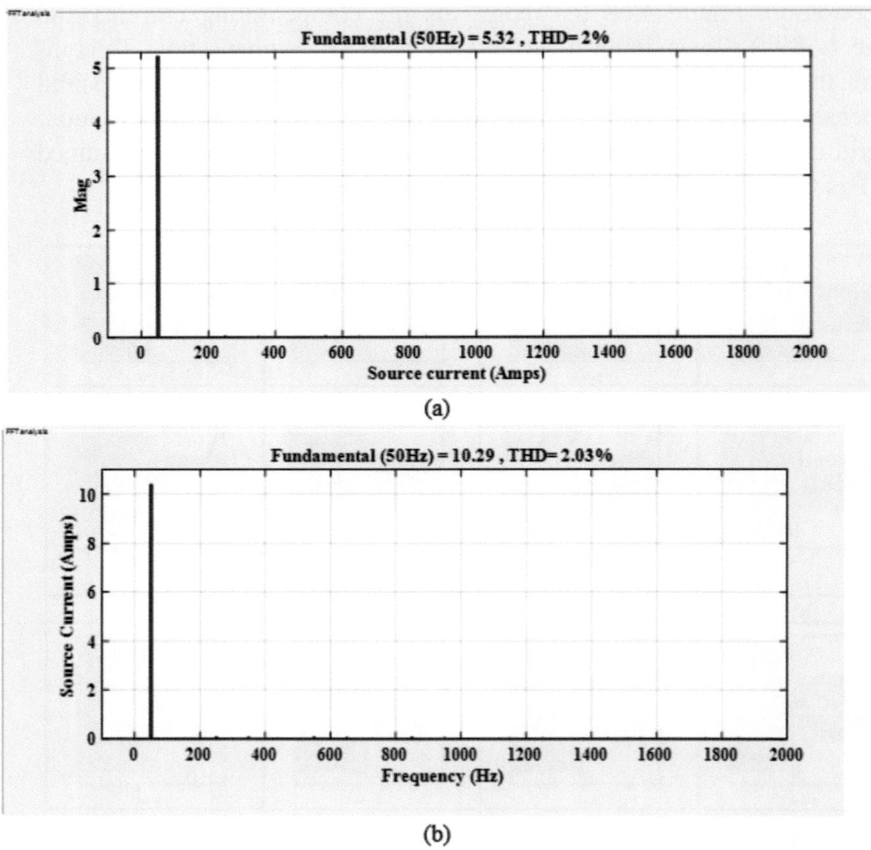

Figure 4.69 THD analysis of double-tuned SHAF (a) steady-state conditions (b) transient-state conditions

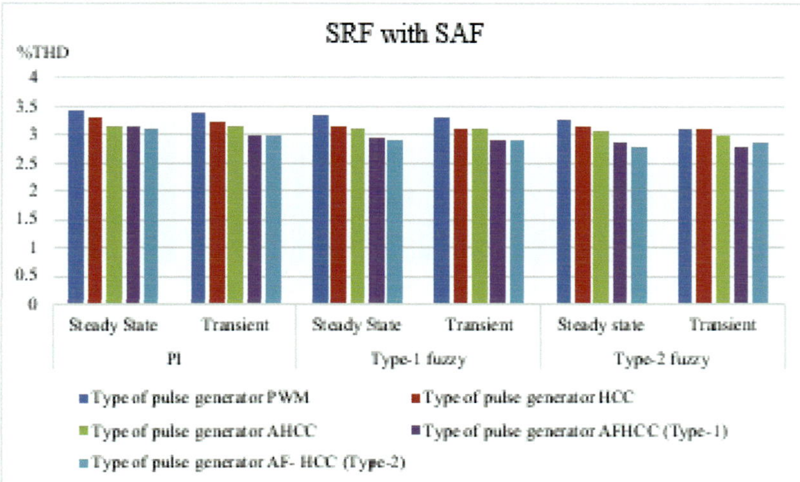

Figure 4.70 Comparison of SAF with various DC voltage controllers and pulse generators.

Table 4.28 THD measurements of SAF with SRF control.

Load	Harmonic con-troller	DC-link con-troller	Condition	Type of pulse generator				
				PWM	HCC	AHCC	AFHCC (type-1)	AF-HCC (type-2)
Diode bridge rectifier with RL load	SRF	PI	Steady State	3.45	3.3	3.15	3.13	3.12
			Transient	3.4	3.21	3.13	3	2.98
		Type-1 fuzzy	Steady State	3.35	3.15	3.12	2.95	2.92
			Transient	3.32	3.12	3.1	2.9	2.9
		Type-2 fuzzy	Steady state	3.27	3.15	3.05	2.85	2.8
			Transient	3.12	3.1	2.97	2.8	2.85

Table 4.28 and Figure 4.70 show the THD measurement of source current with only SAF. The performance of SAF is analyzed with SRF power control as a reference current controller. DC-link controller (PI, type-1 fuzzy, and type-2 fuzzy), and pulse generators (PWM, HCC, AHCC, AFHCC type-1, and AFHCC type-2). Similarly, Table 4.29 and Figure 4.71 show the THD measurement of source current when SAF with two single-tuned passive filters are connected to the system and Table 4.30 and Figure 4.72 show the THD measurement of source current when SAF with double-tuned passive filter is connected.

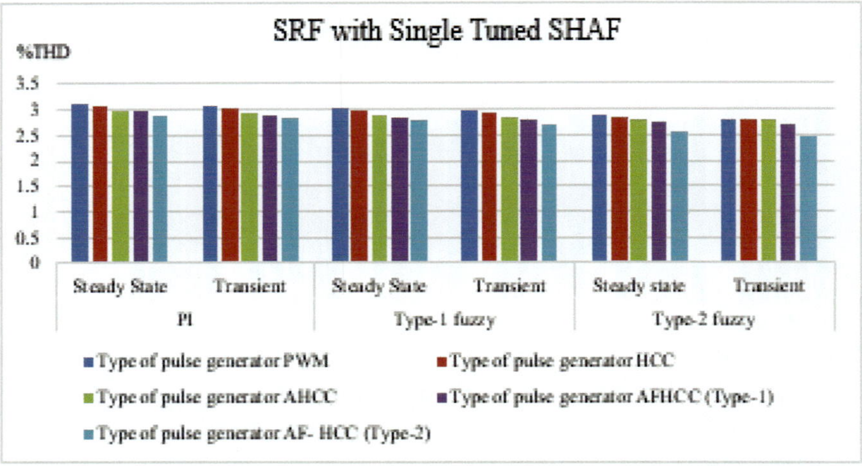

Figure 4.71 Comparison of SAF with various DC voltage controllers and pulse generators.

Table 4.29 THD measurements of single-tuned SHAF with SRF control.

Load	Harmonic controller	DC-link controller	Condition	Type of pulse generator				
				PWM	HCC	AHCC	AFHCC (type-1)	AF-HCC (type-2)
Diode bridge rectifier with RL load	SRF	PI	Steady State	3.11	3.05	3	2.98	2.9
			Transient	3.07	3.02	2.95	2.9	2.85
		Type-1 fuzzy	Steady State	3.01	2.98	2.9	2.82	2.8
			Transient	2.98	2.94	2.85	2.8	2.7
		Type-2 fuzzy	Steady state	2.9	2.85	2.8	2.73	2.55
			Transient	2.8	2.78	2.78	2.7	2.45

Real and Reactive power analysis: The double-tuned SHAF improves the power factor in distribution network. Figure 4.73 shows the waveform of power factor correction for A-phase voltage and current. The real and reactive powers are calculated as given in Tables 4.28 to 4.30. This result indicates that the SHAF is diminishing reactive power and therefore improving the power factor.

Tables 4.31 to 4.33 give information regarding the real and reactive power requirements when connected to shunt active filter, single-tuned SHAF, and double-tuned SHAF to the non-linear system under steady-state conditions.

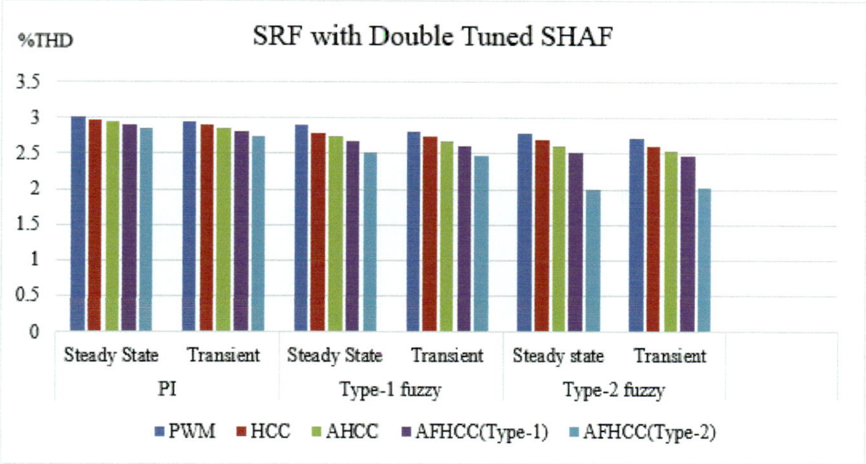

Figure 4.72 Comparison of double-tuned SHAF with various DC voltage controllers and pulse generators.

Table 4.30 THD measurements of double-tuned SHAF with SRF control.

Load	**Harmonic con-troller**	**DC-link con-troller**	**Condition**	**Type of pulse generator**				
				PWM	**HCC**	**AHCC**	**AFHCC (type-1)**	**AF-HCC (type-2)**
Diode bridge rectifier with RL load	SRF	PI	Steady State	3.01	2.98	2.95	2.9	2.85
			Transient	2.95	2.9	2.87	2.82	2.74
		Type-1 fuzzy	Steady State	2.9	2.8	2.74	2.68	2.52
			Transient	2.82	2.75	2.68	2.62	2.48
		Type-2 fuzzy	Steady state	2.8	2.7	2.6	2.51	**2**
			Transient	2.72	2.62	2.55	2.48	**2.03**

Figure 4.74 shows the DC-link voltage across the VSI capacitor bank of SHAF. In Figure 4.63, the SHAF utilizes Type-2 fuzzy logic DC-link voltage controller, AFHCC type-2 pulse generator, and double-tuned passive filter. In a similar manner, all the DC-link control techniques of SAF are able to maintain the constant DC-link voltage across VSI capacitor bank under steady- and transient-state conditions.

Figure 4.75 shows the reactive power balance, power factor correction, and real power demand under steady- and transient-state condition using SHAF. In Figure 4.75, the SHAF utilizes type-2 fuzzy logic DC-link voltage controller, AFHCC type-2 pulse generator, and double-tuned passive filter. In

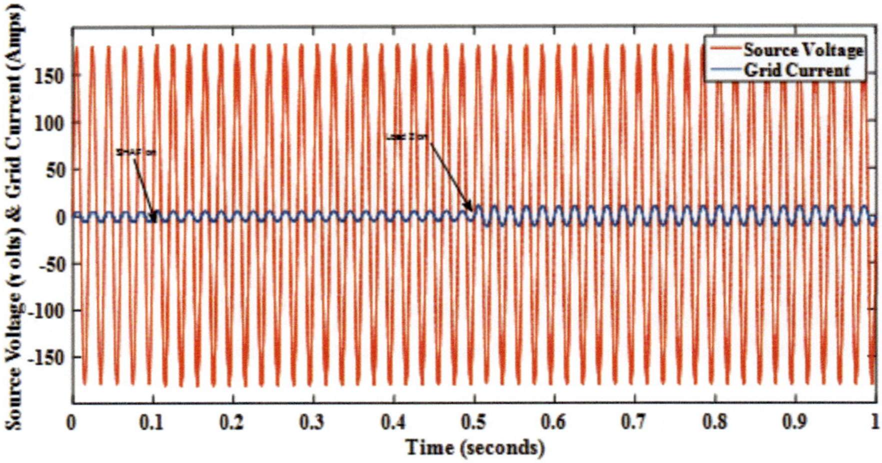

Figure 4.73 Power factor correction by the SHAF.

Table 4.31 Real and reactive powers of SAF with SRF control under steady state.

Load	Reference controller	Without SAF	DC-link controller	With SAF (steady state)				
				PWM	**HCC**	**AHCC**	**AFHCC type-1**	**AFHCC type-2**
Diode bridge rectifier with RL load	SRF	P = 1.5 KW Q = 720 VAR	PI	$P = 1.46$ KW $Q = 60$ VAR	$P = 1.42$ KW $Q = 29$ VAR	$P = 1.43$ KW $Q = 49$ VAR	$P = 1.49$ KW $Q = 25.5$ VAR	$P = 1.43$ KW $Q = 43$ VAR
			Type-1 fuzzy	$P = 1.45$ KW $Q = 100$ VAR	$P = 1.46$ KW $Q = 29$ VAR	$P = 1.48$ KW $Q = 32$ VAR	$P = 1.45$ KW $Q = 50$ VAR	$P = 1.48$ KW $Q =$ 50 VAR
			Type-2 fuzzy	$P = 1.45$ KW $Q = 100$ VAR	$P = 1.46$ KW $Q = 29$ VAR	$P = 1.48$ KW $Q = 32$ VAR	$P = 1.45$ KW $Q =50$ VAR	$P = 1.49$ KW $Q = 50$ VAR

Table 4.32 Real and reactive powers of single-tuned SHAF with SRF control under steady state.

Load	Reference controller	Without SAF	DC-link controller	With single-tuned SHAF (steady state)				
				PWM	**HCC**	**AHCC**	**AFHCC type-1**	**AFHCC type-2**
Diode bridge rectifier with RL load	SRF	P = 1.5 KW Q = 720 VAR	PI	$P = 1.46$ KW $Q = 60$ VAR	$P = 1.42$ KW $Q = 29$ VAR	$P = 1.43$ KW $Q = 49$ VAR	$P = 1.49$ KW $Q = 25.5$ VAR	$P = 1.43$ KW $Q = 43$ VAR
			Type-1 fuzzy	$P = 1.45$ KW $Q = 100$ VAR	$P = 1.46$ KW $Q = 29$ VAR	$P = 1.48$ KW $Q = 32$ VAR	$P = 1.45$ KW $Q = 50$ VAR	$P = 1.48$ KW $Q = 50$ VAR
			Type-2 fuzzy	$P = 1.45$ KW $Q = 100$ VAR	$P = 1.46$ KW $Q = 29$ VAR	$P = 1.48$ KW $Q =$ 32 VAR	$P = 1.45$ KW $Q = 50$ VAR	$P = 1.49$ KW $Q = 50$ VAR

Table 4.33 Real and reactive powers of double-tuned SHAF with SRF control under steady state.

Load	Reference controller	Without SAF	DC-link controller	With double-tuned SHAF (steady state)				
				PWM	HCC	AHCC	AFHCC type-1	AFHCC type-2
Diode bridge rectifier with RL load	SRF	P = 1.5 KW Q = 720 VAR	PI	$P = 1.46$ KW $Q = 60$ VAR	$P = 1.45$ KW $Q =$ 29 VAR	$P = 1.43$ KW $Q =$ 49 VAR	$P = 1.49$ KW $Q =$ 25.5 VAR	$P = 1.45$ KW $Q =$ 43 VAR
			Type-1 fuzzy	$P = 1.45$ KW $Q =$ 100 VAR	$P = 1.46$ KW $Q =$ 29 VAR	$P = 1.48$ KW $Q = 32$ VAR	$P = 1.45$ KW $Q =$ 50 VAR	$P = 1.48$ KW $Q = 50$ VAR
			Type-2 fuzzy	$P = 1.45$ KW $Q = 100$ VAR	$P = 1.46$ KW $Q = 29$ VAR	$P = 1.48$ KW $Q = 32$ VAR	$P = 1.45$ KW $Q = 50$ VAR	$P = 1.49$ KW $Q = 50$ VAR

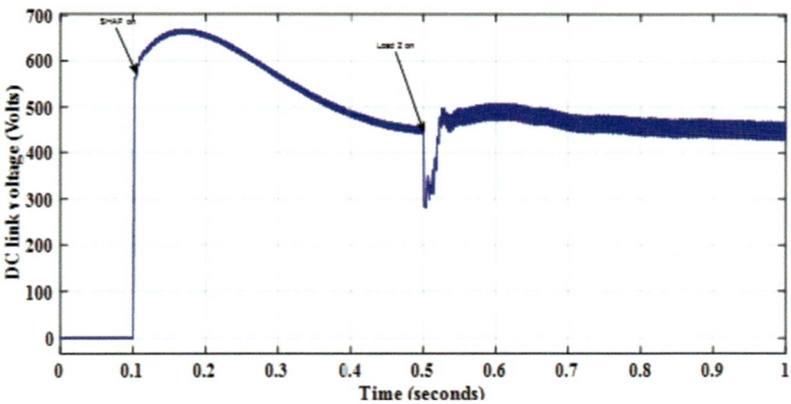

Figure 4.74 DC-link voltage across the VSI capacitor.

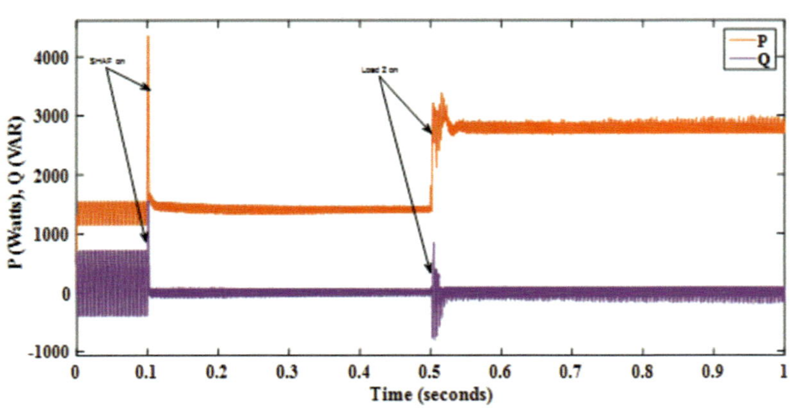

Figure 4.75 Real power and reactive power demands.

a similar manner, the oscillating component of reactive power, real power, and power factor has been eliminated by all the proposed SAF, single-tuned, and double-tuned SHAF.

Remarks: The SRF theory with various PWM pulse generators-based SHAF is simulated and compensation principle is evaluated. The AFHCC type-2 reduces the switching power losses and improves the SHAF performance than other techniques. From these results, it is proved that the THD of the grid current follows harmonic standards IEEE 519-2014.

4.7 Modified SRF Theory

The modified SRF theory with different PWM pulse generators, DC capacitor voltage regulators, and tuned passive filters is performed by using MATLAB/Simulink tools. The model system is having three-phase VSI, coupling inductor, time-domain reference current extraction controller based on modified SRF theory, pulse generator, and tuned passive filters as represented in Figure 4.76. It is used to pull out the desired reference current from the polluted load current. The reference current is related with original compensation current to produce the trigger pulses for SAF using PWM/HCC/AHCC/AFHCC type-1/AFHCC type-2 controllers. The SAF generates the anti-harmonic current to cancel out current harmonics and balance reactive power in the distributed network. The whole system is tested under steady- and transient-state conditions. Further, the tuned passive

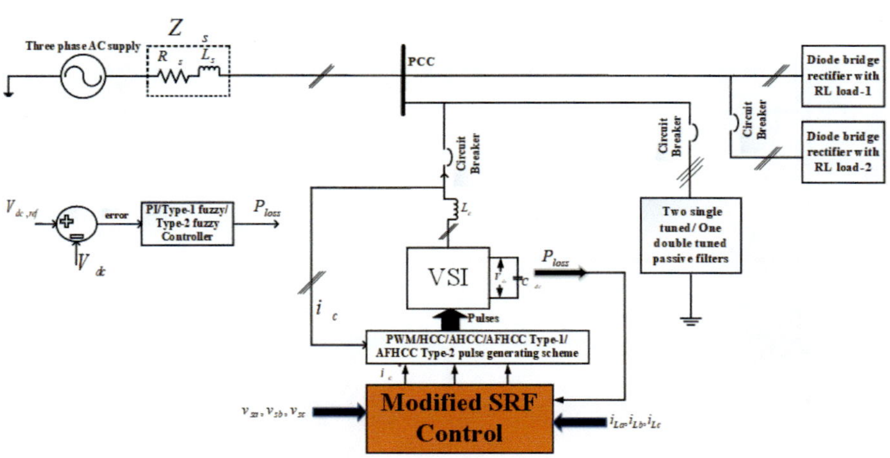

Figure 4.76 Modified SRF theory-based SHAF.

filters like single- and double-tuned combine with the SAF giving better performance.

4.7.1 Case-1 – PI DC-link voltage controller

PI DC-link voltage controller with different pulse generators is simulated for SHAF under balanced uncontrolled diode bridge rectifiers with RL-load conditions. The waveforms in Figure 4.77 are concentrated on PI with PWM/HCC/AHCC/AFHCC type-1 technique respectively. Figure 4.77 (a) shows the load current without SHAF. It is observed that the load current is having power frequency as well as harmonics due to the non-linear load (diode bridge rectifier load). The PWM/HCC/AHCC/AFHCC type-1/AFHCC type-2 pulse-generating techniques have been used to drive the SAF. The SHAF when switched on at 0.1 sec, provides the desired compensating current to cancel out the harmonic current at PCC as shown in Figure 4.77 (b). By injecting equal but opposite harmonic components, the harmonic compensation has been achieved. Figure 4.77 (c) shows the grid current after harmonic compensation. These waveforms are achieved under

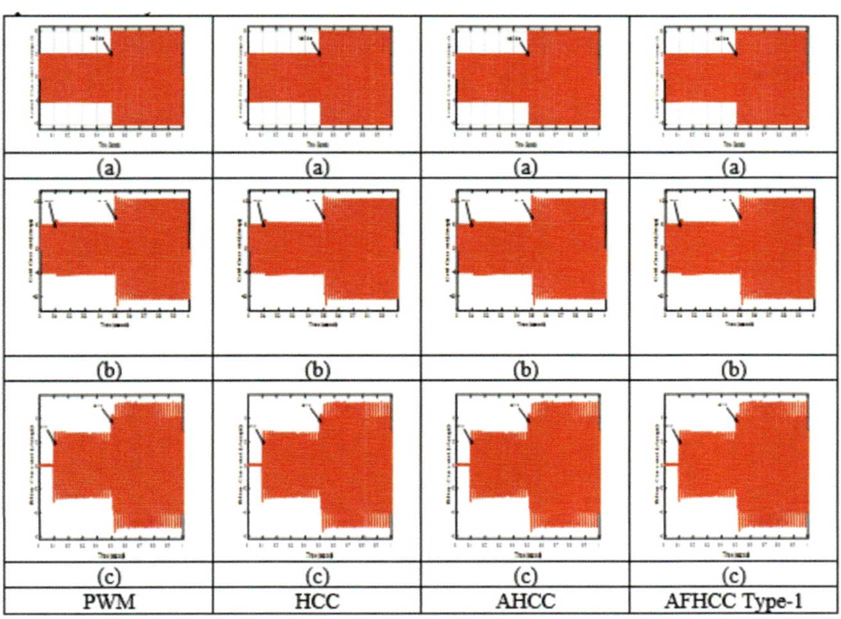

Figure 4.77 Simulation waveforms of (a) source current before compensation, (b) compensation currents, and (c) source current after compensation.

uncontrolled bridge rectifier load with steady and transient conditions and these are for A-phase alone; B and C-phase are phase shifted by 120°.

Figure 4.78 shows the grid current of phase A, when SAF with PI DC-link voltage controller and AFHCC type-1 switching scheme is connected to double-tuned passive filter at 0.1 sec.

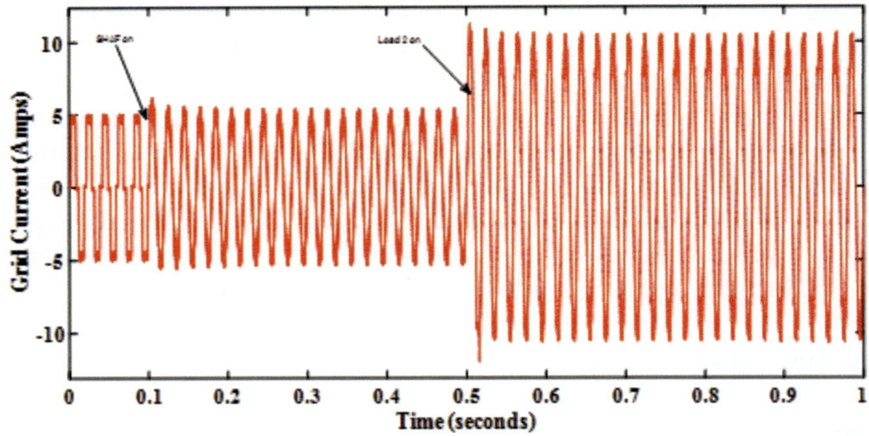

Figure 4.78 Double-tuned SHAF grid currents with PI DC-link voltage controller and HCC switching scheme.

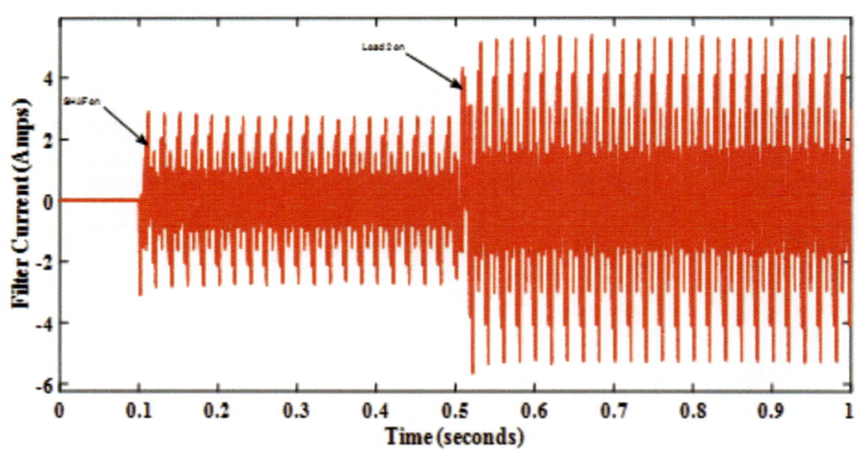

Figure 4.79 Double-tuned SHAF compensating currents with PI DC-link voltage.

Figure 4.79 shows the SAF compensation currents with type-1 fuzzy logic DC-link voltage controller and AFHCC type-1 switching scheme and double-tuned passive filter.

4.7.2 Case-2 – Type-1 fuzzy logic DC-link voltage controller

Type-1 fuzzy logic DC-link voltage controllers with PWM/HCC/AHCC /AFHCC type-1/AFHCC type-2 are simulated under uncontrolled bridge rectifier with RL load. Figure 4.80 is achieved by using PWM/HCC/AHCC /AFHCC type-1 under uncontrolled diode bridge rectifier load with steady and transient state. Figure 4.80 (a) shows the source currents without compensation. Figure 4.80 (b) shows the compensating currents and Figure 4.80 (c) shows the grid currents after compensation. To mitigate the harmonic currents, the double-tuned SHAF (when it is ON) has to provide the desired compensating currents as shown in Figure 4.80 (b). The grid currents after

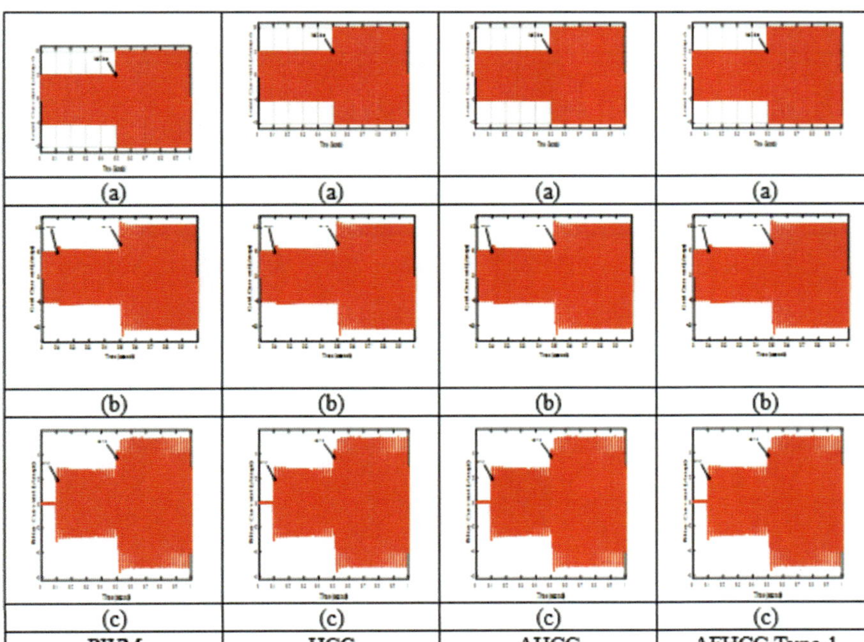

Figure 4.80 Simulation waveforms of (a) source current before compensation, (b) compensation currents, and (c) source current after compensation.

compensation are shown in Figure 4.80 (c), the double-tuned SHAF is ON, the source currents become sinusoidal.

4.7.3 Case-3 – Type-2 fuzzy logic controller

Type-2 fuzzy logic DC-link voltage controllers with PWM/HCC/AHCC /AFHCC type-2/AFHCC type-2 pulse generators are simulated under uncontrolled diode bridge rectifier with RL load. Figure 4.81 is achieved using PWM/HCC/AHCC/AFHCC type-1 under uncontrolled diode bridge rectifier load with steady and transient state. Figure 4.81 (a) shows the grid currents without compensation. Figure 4.81 (b) represents the compensating currents and Figure 4.81 (c) shows the grid currents with SHAF compensation. To compensate the harmonic currents, the double-tuned SHAF has to give the desired compensating currents as represented in Figure 4.81 (b). After compensation, the grid currents are shown in Figure 4.81 (c), i.e., when the double-tuned SHAF is ON, the grid currents become mostly sinusoidal.

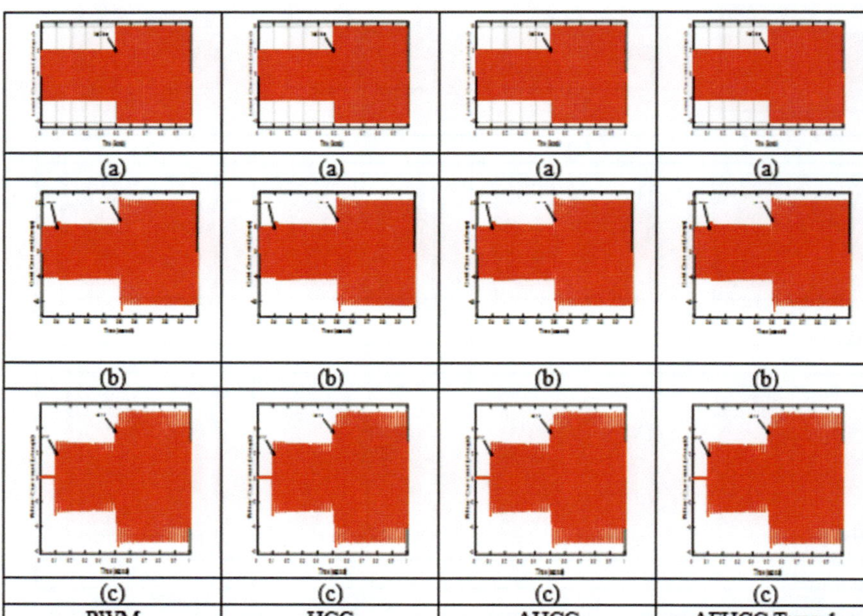

Figure 4.81 Simulation waveforms of (a) source current before compensation, (b) compensation currents, and (c) source current after compensation.

THD analysis: The harmonics against magnitude of the grid current is plotted with double-tuned SHAF under uncontrolled bridge rectifier with RL load. The simulation results are shown in Figure 4.82 is for type-2 fuzzy DC capacitor voltage controller and AFHCC type-2 pulse generator-based double-tuned SHAF under steady- and transient-state condition with diode bridge rectifier. It has been observed that the grid current without SHAF is having harmonic components whereas with double-tuned SHAF the THD fell from 26.12% to 1.84% in steady state and from 26.21% to 1.94% in transient state respectively as represented in Figure 4.82.

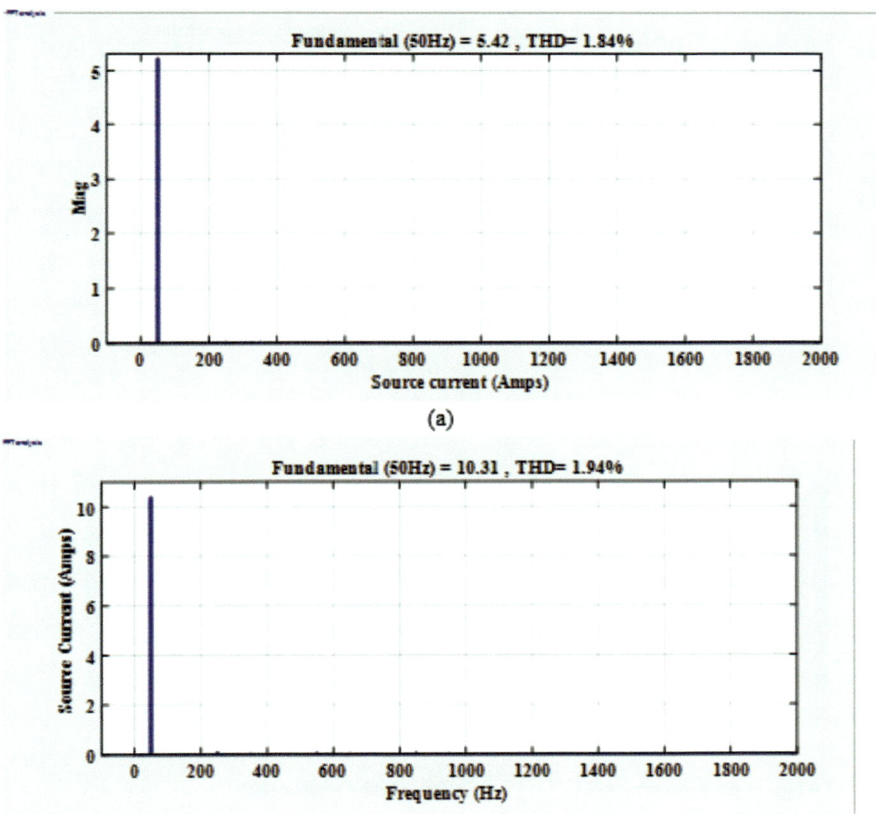

(b)

Figure 4.82 THD analysis of double-tuned SHAF (a) steady-state conditions (b) transient-state conditions.

Table 4.34 and Figure 4.83 show the THD measurement of source current with only SAF. The performance of SAF is analyzed with modified SRF power control as a reference current controller, DC-link controller (PI, type-1 fuzzy, and type-2 fuzzy), and pulse generators (PWM, HCC, AHCC, AFHCC type-1, and AFHCC type-2). Similarly, Table 4.35 and Figure 4.84 show the THD measurement of source current when SAF with two single-tuned passive filters are connected to the system and Table 4.36 and Figure 4.85 show the THD measurement of source current when SAF with double-tuned passive filter is connected.

Table 4.34 THD measurements of SAF with modified SRF control

Load	Harmonic controller	DC-link controller	Condition	Type of pulse generator				
				PWM	HCC	AHCC	AFHCC type-1	AF-HCC type-2
Diode bridge rectifier with RL load	Modified SRF	PI	Steady State	3.35	3.2	3.1	3	2.98
			Transient	3.3	3.15	3.05	2.96	2.85
		Type-1 fuzzy	Steady State	3.27	3.1	3	2.9	2.8
			Transient	3.2	3.05	2.96	2.88	2.78
		Type-2 fuzzy	Steady state	3.16	2.98	2.9	2.82	2.6
			Transient	3.1	2.92	2.9	2.72	2.62

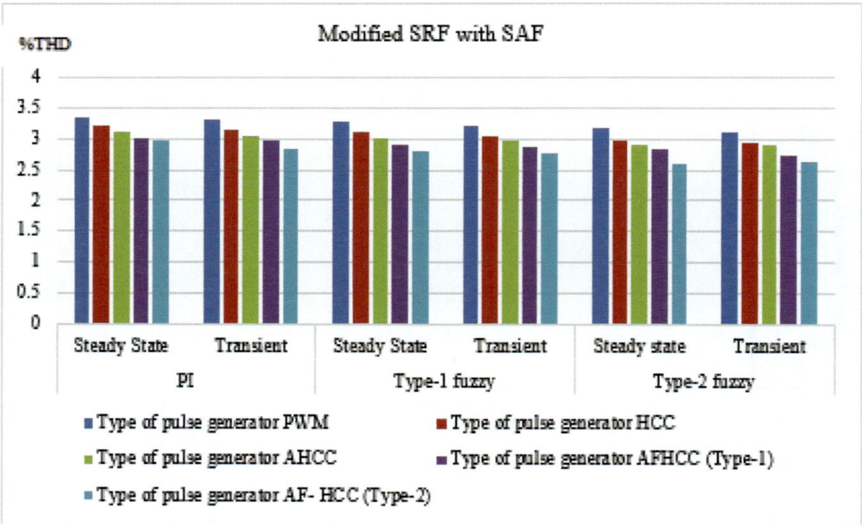

Figure 4.83 Comparison of SAF with various DC voltage controllers and pulse generators.

Table 4.35 THD measurements of single-tuned SHAF with modified SRF control.

Load	Harmonic controller	DC-link controller	Condition	Type of pulse generator				
				PWM	HCC	AHCC	AFHCC type-1	AF-HCC type-2
Diode bridge rectifier with RL load	Modified SRF	PI	Steady State	3.01	2.95	2.9	2.85	2.78
			Transient	2.98	2.92	2.85	2.82	2.72
		Type-1 fuzzy	Steady State	2.9	2.88	2.8	2.75	2.65
			Transient	2.85	2.82	2.75	2.68	2.61
		Type-2 fuzzy	Steady state	2.8	2.78	2.7	2.65	2.4
			Transient	2.73	2.65	2.68	2.62	2.35

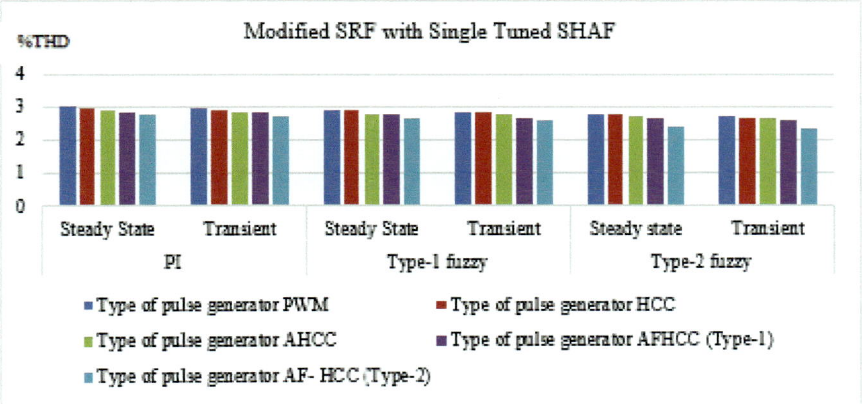

Figure 4.84 Comparison of single-tuned SHAF with various DC voltage controllers and pulse generators.

Table 4.36 THD measurements of double-tuned SHAF with modified SRF control.

Load	Harmonic controller	DC-link controller	Condition	Type of pulse generator				
				PWM	HCC	AHCC	AFHCC type-1	AF-HCC type-2
Diode bridge rectifier with RL load	Modified SRF	PI	Steady State	2.9	2.72	2.61	2.58	2.44
			Transient	2.85	2.7	2.58	2.49	2.32
		Type-1 fuzzy	Steady State	2.8	2.75	2.54	2.43	2.28
			Transient	2.75	2.7	2.52	2.35	2.1
		Type-2 fuzzy	Steady state	2.7	2.63	2.48	2.28	1.84
			Transient	2.65	2.57	2.43	2.2	1.94

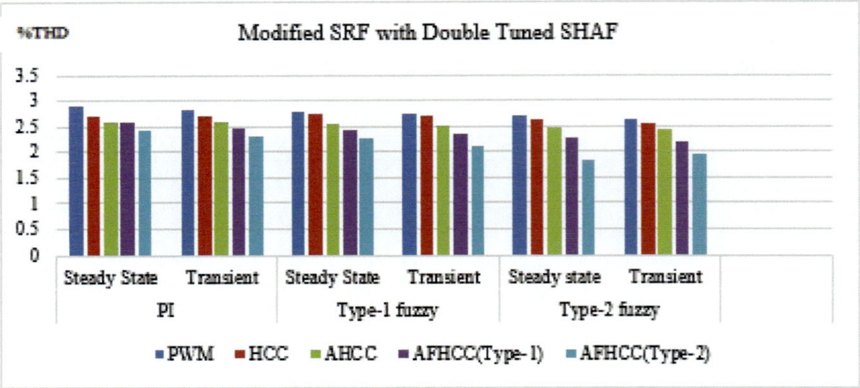

Figure 4.85 Comparison of single-tuned SHAF with various DC voltage controllers and pulse generators.

Real and Reactive power analysis: The double-tuned SHAF improves the power factor in distribution network. Figure 4.86 shows the waveform of power factor correction for A-phase voltage and current. The real and reactive powers are calculated as given in Tables to 4.39. This result indicates that the SHAF is diminishing reactive power and therefore improving the power factor.

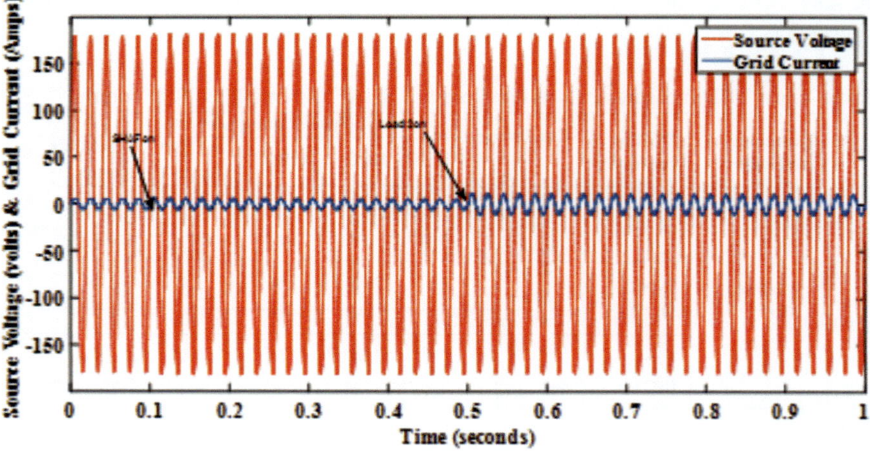

Figure 4.86 Power factor correction by the SHAF.

Tables 4.37 to 4.39 give the information regarding the real and reactive power requirements when connected to shunt active filter, single-tuned SHAF, and double-tuned SHAF to the non-linear system under steady-state conditions.

Figure 4.87 shows the DC-link voltage across the VSI capacitor bank of SHAF. In Figure 4.87, the SHAF utilizes type-2 fuzzy logic DC-link voltage controller, AFHCC type-2 pulse generator, and double-tuned passive filter. In a similar manner, all the DC-link control techniques of SAF are able to maintain the constant DC-link voltage across VSI capacitor bank under steady- and transient-state conditions.

Figure 4.88 shows the reactive power balance, power factor correction, and real power demand under steady- and transient-state conditions using SHAF. In Figure 4.87, the SHAF utilizes type-2 fuzzy logic DC-link voltage controller, AFHCC type-2 pulse generator, and double-tuned passive filter. In

Table 4.37 Real and reactive powers of SAF with modified SRF control under steady state.

Load	Reference controller	Without SAF	DC-link controller	With SHAF (steady state)				
				PWM	HCC	AHCC	AFHCC type-1	AFHCC type-2
Diode bridge rectifier with RL load	Modified SRF	P = 1.5 KW Q = 720 VAR	PI	$P = 1.46$ KW $Q = 60$ VAR	$P = 1.42$ KW $Q = 29$ VAR	$P = 1.43$ KW $Q = 49$ VAR	$P = 1.49$ KW $Q = 25.5$ VAR	$P = 1.43$ KW $Q = 43$ VAR
			Type-1 fuzzy	$P = 1.45$ KW $Q = 100$ VAR	$P = 1.46$ KW $Q =$ 29 VAR	$P = 1.48$ KW $Q = 32$ VAR	$P = 1.45$ KW $Q = 50$ VAR	$P = 1.48$ KW $Q = 50$ VAR
			Type-2 fuzzy	$P = 1.45$ KW $Q = 100$ VAR	$P = 1.46$ KW $Q = 29$ VAR	$P = 1.48$ KW $Q = 32$ VAR	$P = 1.55$ KW $Q =50$ VAR	$P = 1.49$ KW $Q = 50$ VAR

Table 4.38 Real and reactive powers of single-tuned SHAF with modified SRF control under steady state.

Load	Reference controller	Without SAF	DC-link controller	With single-tuned SHAF (steady state)				
				PWM	HCC	AHCC	AFHCC type-1	AFHCC type-2
Diode bridge rectifier with RL load	Modified SRF	P = 1.5 KW Q = 720 VAR	PI	$P = 1.46$ KW $Q = 60$ VAR	$P = 1.42$ KW $Q = 29$ VAR	$P = 1.45$ KW $Q = 49$ VAR	$P = 1.49$ KW $Q = 25.5$ VAR	$P = 1.43$ KW $Q = 43$ VAR
			Type-1 fuzzy	$P = 1.45$ KW $Q = 100$ VAR	$P = 1.46$ KW $Q = 29$ VAR	$P = 1.48$ KW $Q = 32$ VAR	$P = 1.45$ KW $Q = 50$ VAR	$P = 1.48$ KW $Q = 50$ VAR
			Type-2 fuzzy	$P = 1.45$ KW $Q = 100$ VAR	$P = 1.46$ KW $Q = 29$ VAR	$P = 1.48$ KW $Q = 32$ VAR	$P = 1.45$ KW $Q = 50$ VAR	$P = 1.49$ KW $Q = 50$ VAR

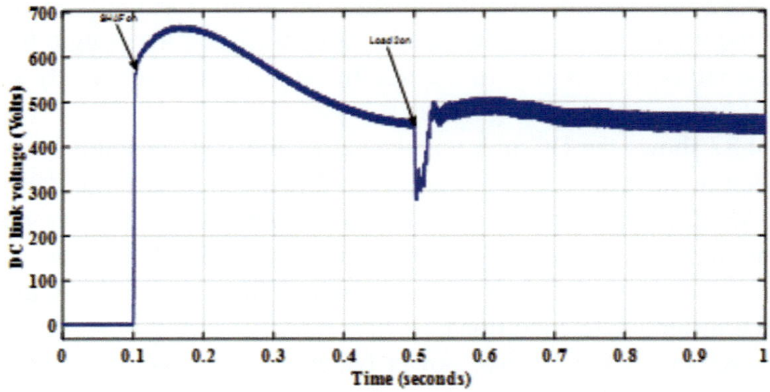

Figure 4.87 DC-link voltage across the VSI capacitor.

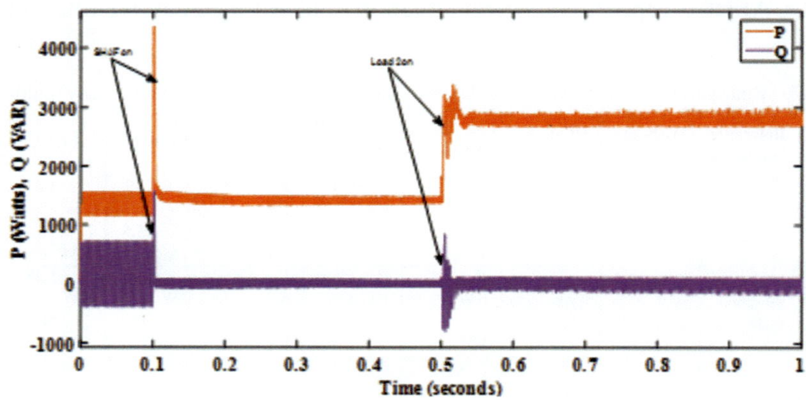

Figure 4.88 Real power, reactive power demands.

Table 4.39 Real and reactive powers of double-tuned SHAF with modified SRF control under steady state.

Load	Reference controller	Without SAF	DC-link controller	With double-tuned SHAF (steady state)				
				PWM	HCC	AHCC	AFHCC type-1	AFHCC type-2
Diode bridge rectifier with RL load	Modified SRF	P = 1.5 KW Q = 720 VAR	PI	$P = 1.46$ KW $Q = 60$ VAR	$P = 1.44$ KW $Q = 29$ VAR	$P = 1.43$ KW $Q = 49$ VAR	$P = 1.49$ KW $Q =$ 25.5 VAR	$P = 1.43$ KW $Q = 43$ VAR
			Type-1 fuzzy	$P = 1.45$ KW $Q = 100$ VAR	$P = 1.46$ KW $Q = 29$ VAR	$P = 1.48$ KW $Q = 32$ VAR	$P = 1.45$ KW $Q = 50$ VAR	$P = 1.48$ KW $Q = 50$ VAR
			Type-2 fuzzy	$P = 1.45$ KW $Q = 100$ VAR	$P = 1.46$ KW $Q = 29$ VAR	$P = 1.48$ KW $Q = 32$ VAR	$P = 1.45$ KW $Q =50$ VAR	$P = 1.49$ KW $Q = 50$ VAR

a similar manner, the oscillating component of reactive power, real power, and power factor has been eliminated by all the proposed SAF, single- and double-tuned SHAF.

Remarks: The modified SRF theory with various PWM pulse generators-based SHAF is simulated and compensation principle is evaluated. The AFHCC type-2 reduces the switching power losses and improves the SHAF performance than other techniques. From these results, it is proved that the THD of the grid current is following harmonic standards IEEE 519-2014.

4.8 Proposed Apq-SRF Control Theory

The proposed Apq-SRF control theory with different PWM pulse generators, DC capacitor voltage regulators, and tuned passive filters is performed by using MATLAB/Simulink tools. The model system is having three-phase VSI, coupling inductor, time-domain reference current extraction controller based on proposed Apq-SRF control theory, pulse generator, and tuned passive filters as represented in Figure 4.89. It is used to pull out the desired reference current from the polluted load current. The reference current is related with original compensation current to produce the trigger pulses for SAF using PWM/HCC/AHCC/AFHCC type-1/AFHCC type-2 controllers. The SAF generates the anti-harmonic current to cancel out current harmonics and balance reactive power in the distributed network. The whole system is tested under steady- and transient-state conditions. Further, the tuned passive

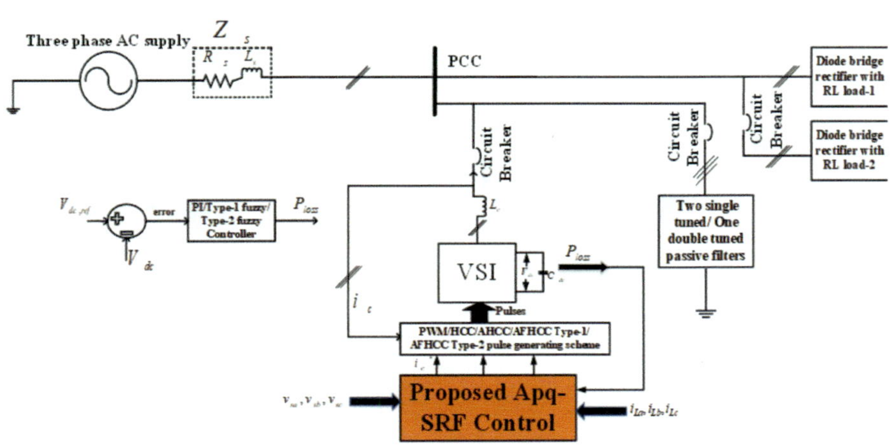

Figure 4.89 Proposed Apq-SRF control-based SHAF.

filters like single- and double-tuned combine with the SAF giving better performance.

4.8.1 Case-1 – PI DC-link voltage controller

PI DC-link voltage controller with different pulse generators is simulated for SHAF under balanced uncontrolled diode bridge rectifiers with RL-load conditions. The waveforms in Figure 4.90 are concentrated on PI with PWM/HCC/AHCC/AFHCC type-2 techniques respectively. Figure 4.90 (a) shows the load current without SHAF. It is observed that the load current is having power frequency as well as harmonics due to the non-linear load (diode bridge rectifier load). The PWM/HCC/AHCC/AFHCC type-1/AFHCC type-2 pulse-generating techniques have been used to drive

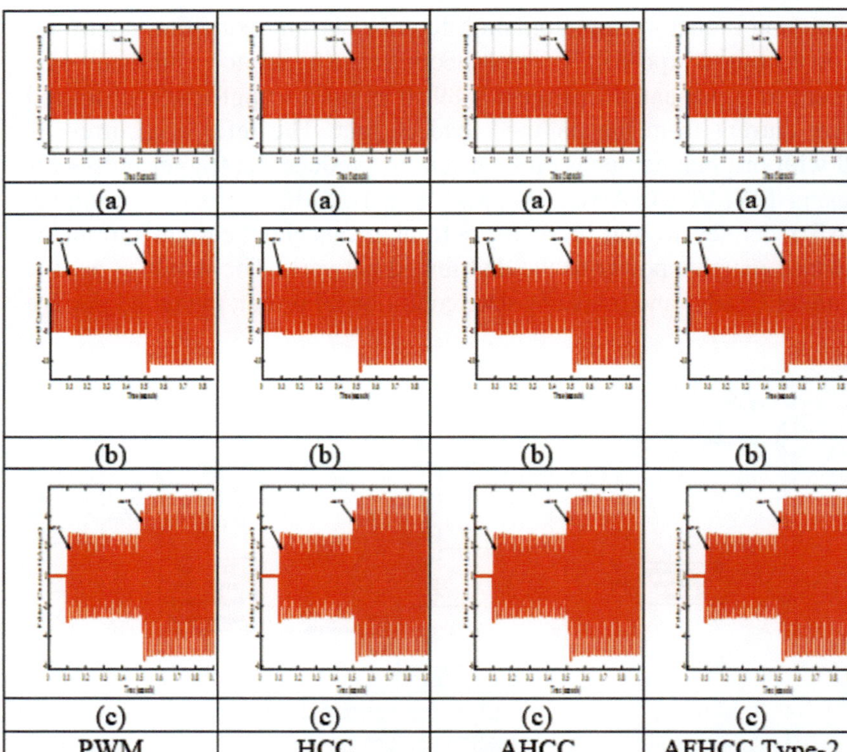

Figure 4.90 Simulation waveforms of (a) source current before compensation, (b) compensation currents, and (c) source current after compensation.

the SAF. The SHAF when switched ON at 0.1 sec provides the desired compensating current to cancel out the harmonic current at PCC as shown in Figure 4.90 (b). By injecting equal but opposite harmonic components, the harmonic compensation has been achieved. Figure 4.90 (c) shows the grid current after harmonic compensation. These waveforms are achieved under uncontrolled bridge rectifier load with steady and transient conditions and these are for A-phase alone; B- and C-phase are phase shifted by 120°. The

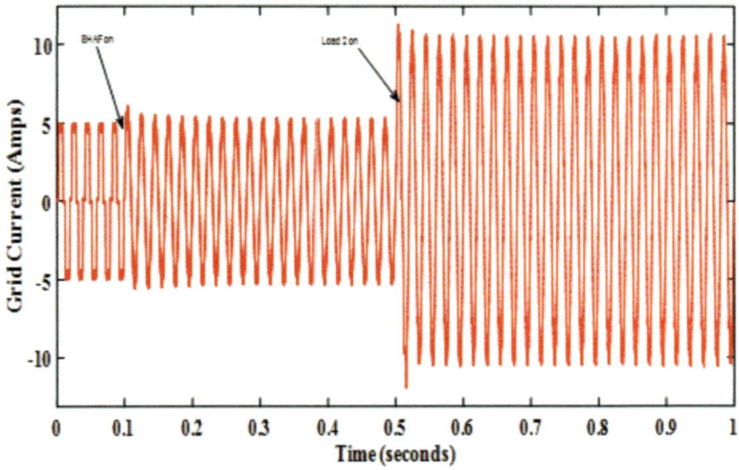

Figure 4.91 Double-tuned SHAF grid currents with PI DC-link voltage controller and HCC switching scheme.

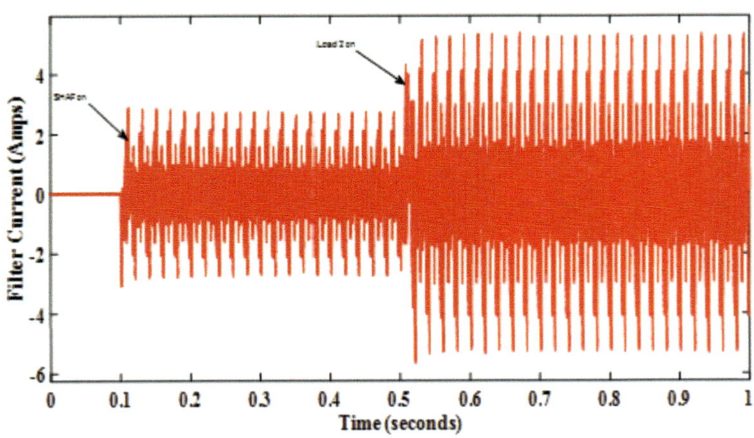

Figure 4.92 Double-tuned SHAF compensating currents with PI DC-link voltage.

THD measurement of source/load current without filter under steady-state condition is 26.12% and under transient condition is 26.01%.

The grid current of phase A, when SAF with PI DC-link voltage controller and HCC switching scheme connected to double-tuned passive filter at 0.1 sec is as shown in Figure 4.91.

The SAF compensation currents with PI DC-link voltage controller, HCC switching scheme, and double-tuned passive filter is shown in Figure 4.92.

4.8.2 Case-2 – Type-1 fuzzy logic DC-link voltage controller

Type-1 fuzzy logic DC-link voltage controllers with PWM/HCC/AHCC /AFHCC type-1/AFHCC type-2 are simulated under uncontrolled bridge rectifier with RL load. Figure 4.93 is achieved by using PWM/HCC/AHCC

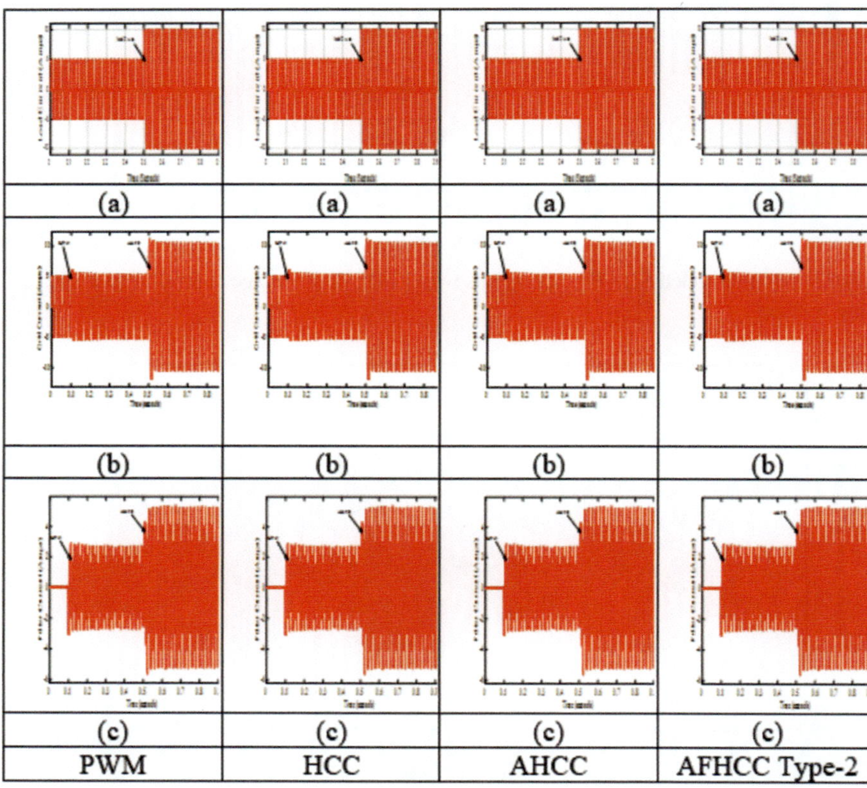

Figure 4.93 Simulation waveforms of (a) source current before compensation, (b) compensation currents, and (c) source current after compensation.

/AFHCC type-2 under uncontrolled diode bridge rectifier load with steady and transient state. Figure 4.93 (a) shows the source currents without compensation. Figure 4.93 (b) shows the compensating currents and Figure 4.93 (c) shows the grid currents after compensation. To mitigate the harmonic currents, the double-tuned SHAF (when it is ON) has to provide the desired compensating currents as shown in Figure 4.93 (b). The grid currents after compensation are shown in Figure 4.93 (c), the double-tuned SHAF is ON, the source currents become sinusoidal.

4.8.3 Case-3 – Type-2 fuzzy logic DC-link voltage controller

Type-2 fuzzy logic DC-link voltage controller with PWM/HCC/AHCC /AFHCC type-1/AFHCC type-2 is simulated under uncontrolled diode bridge rectifier with RL load. Figure 4 94 is achieved using PWM/HCC/AHCC

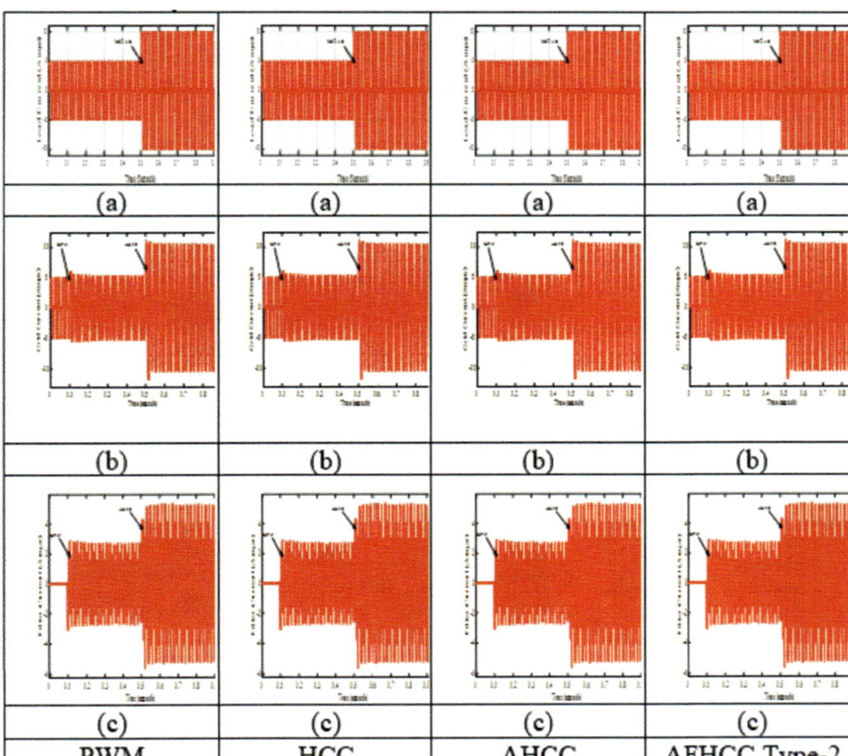

Figure 4.94 Simulation waveforms of (a) source current before compensation, (b) compensation currents, and (c) source current after compensation.

/AFHCC type-2 under uncontrolled diode bridge rectifier load with steady and transient state. Figure 4.94 (a) shows the grid currents without compensation. Figure 4.94 (b) represents the compensating currents and Figure 4.94 (c) shows the grid currents with SHAF compensation. To compensate the harmonic currents, the double-tuned SHAF has to give the desired compensating currents as represented in Figure 4.94 (b). After compensation, the grid currents are shown in Figure 4.94 (c), i.e., when the double-tuned SHAF is ON, the grid currents become mostly sinusoidal.

THD analysis: The harmonics against magnitude of the grid current is plotted with double-tuned SHAF under uncontrolled bridge rectifier with RL load. The simulation results are shown in Figure 4.95 is for type-2 fuzzy DC capacitor voltage controller and AFHCC type-2 pulse generator based-double-tuned SHAF under steady- and transient-state conditions with diode bridge rectifier. It has been observed that the grid current without SHAF is having harmonic components whereas with double-tuned SHAF the THD fell from 26.12% to 1.84% in steady state and 26.21% to 1.94% in transient state respectively as represented in Figure 4.95.

Table 4.40 THD measurements of SAF with proposed Apq-SRF control.

Load	Harmonic controller	DC-link controller	Condition	Type of pulse generator				
				PWM	HCC	AHCC	AFHCC type-1	AF-HCC type-2
Diode bridge rectifier with RL load	Proposed Apq-SRF	PI	Steady State	3.27	3.18	3.09	2.91	2.85
			Transient	3.23	3.1	2.98	2.89	2.82
		Type-1 fuzzy	Steady State	3.2	3.05	2.9	2.82	2.78
			Transient	3.15	3	2.85	2.8	2.72
		Type-2 fuzzy	Steady state	3.1	2.9	2.8	2.75	2.55
			Transient	3.05	2.85	2.78	2.7	2.52

Table 4.41 THD measurements of single-tuned SHAF with proposed Apq-SRF control.

Load	Harmonic controller	DC-link controller	Condition	Type of pulse generator				
				PWM	HCC	AHCC	AFHCC type-1	AF-HCC type-2
Diode bridge rectifier with RL load	Proposed Apq-SRF	PI	Steady State	2.95	2.9	2.85	2.8	2.73
			Transient	2.88	2.82	2.78	2.75	2.7
		Type-1 fuzzy	Steady State	2.8	2.78	2.7	2.65	2.65
			Transient	2.78	2.72	2.68	2.62	2.58
		Type-2 fuzzy	Steady state	2.7	2.68	2.62	2.59	2.38
			Transient	2.65	2.61	2.58	2.52	2.31

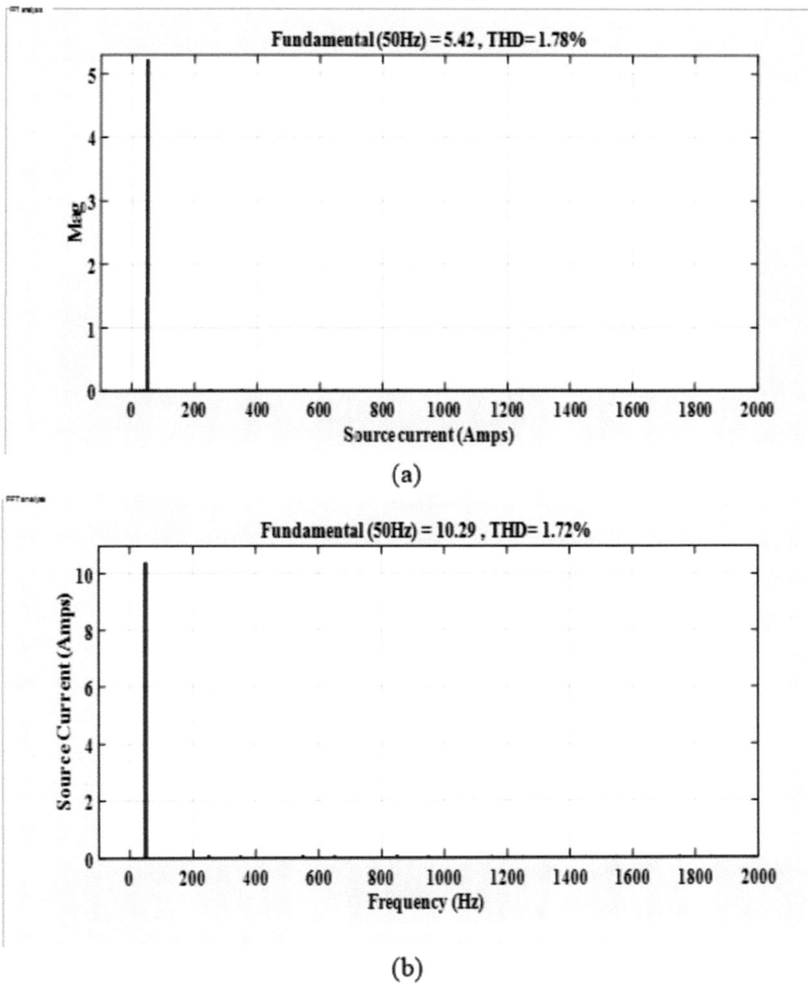

Figure 4.95 THD analysis of double-tuned SHAF (a) steady-state conditions (b) transient-state conditions.

Table 4.40 and Figure 4.96 show the THD measurement of source current with only SAF. The performance of SAF is analyzed with proposed Apq-SRF power control as a reference current controller, DC-link controller (PI, type-1 fuzzy, and type-2 fuzzy), and pulse generators (PWM, HCC, AHCC, AFHCC type-1, and AFHCC type-2). Similarly, Table 4.41 and Figure 4.97 show the THD measurement of source current when SAF with two single-tuned passive filters are connected to the system and Table 4.42 and Figure 4.97 show the THD measurement of source current when SAF with double-tuned passive filter is connected.

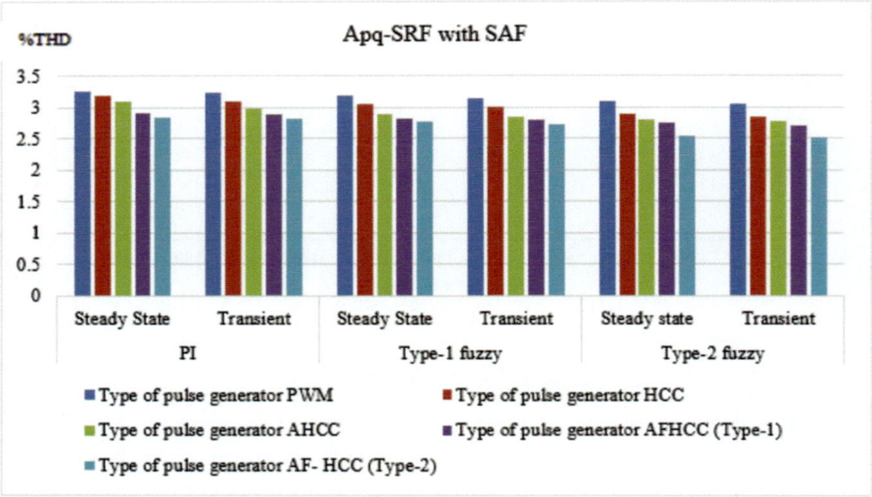

Figure 4.96 Comparison of SAF with various DC voltage controllers and pulse generators.

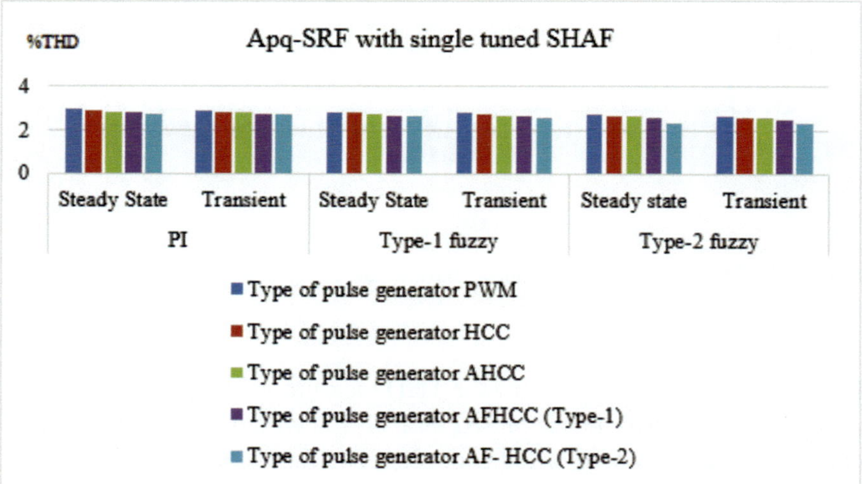

Figure 4.97 Comparison of single-tuned SHAF with various DC voltage controllers and pulse generators.

Table 4.42 THD measurements of double-tuned SHAF with proposed Apq-SRF control.

Load	Harmonic controller	DC-link controller	Condition	Type of pulse generator				
				PWM	HCC	AHCC	AFHCC type-1	AF-HCC type-2
Diode bridge rectifier with RL load	Proposed Apq-SRF	PI	Steady State	2.8	2.72	2.68	2.58	2.44
			Transient	2.78	2.7	2.58	2.49	2.32
		Type-1 fuzzy	Steady State	2.7	2.65	2.54	2.43	2.28
			Transient	2.68	2.6	2.52	2.35	2.1
		Type-2 fuzzy	Steady state	2.6	2.55	2.48	2.28	1.78
			Transient	2.58	2.5	2.43	2.2	1.72

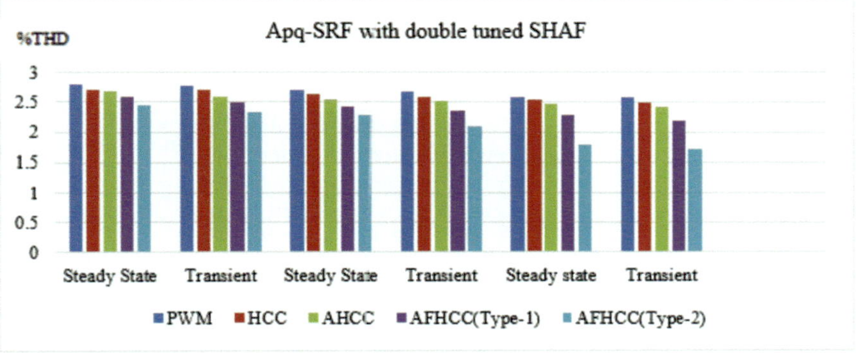

Figure 4.98 Comparison of double-tuned SHAF with various DC voltage controllers and pulse generators.

Real and Reactive power analysis: The double-tuned SHAF improves the power factor in distribution network. Figure 4.99 shows the waveform of power factor correction for A-phase voltage and current. The real and reactive powers are calculated as given in Tables 4.40 to 4.42. This result indicates that the SHAF is diminishing reactive power and therefore improving the power factor.

Tables 4.43 to 4.45 give the information regarding the real and reactive power requirements when connected to shunt active filter, single-tuned SHAF, and double-tuned SHAF to the non-linear system under transient-state condition.

Figure 4.100 shows the DC-link voltage across the VSI capacitor bank of SHAF. In Figure 4.98, the SHAF utilizes type-2 fuzzy logic DC-link voltage controller, AFHCC type-2 pulse generator, and double-tuned passive filter. In a similar manner, all the DC-link control techniques of SAF are able

Table 4.43 Real and reactive powers of SAF with proposed Apq-SRF control under transient state.

Load	Reference controller	Without SAF	DC-link controller	With SAF (transient state)				
				PWM	HCC	AHCC	AFHCC type-1	AFHCC type-2
Diode bridge rectifier with RL load	Proposed Apq-SRF	$P = 3$ KW $Q = 1540$ VAR	PI	$P = 2.8$ KW $Q = 90$ VAR	$P = 2.81$ KW $Q = 92$ VAR	$P = 2.78$ KW $Q = 91$ VAR	$P = 2.75$ KW $Q = 92$ VAR	$P = 2.79$ KW $Q = 91$ VAR
			Type-1 fuzzy	$P = 2.8$ KW $Q = 90$ VAR	$P = 2.81$ KW $Q = 92$ VAR	$P = 2.78$ KW $Q = 91$ VAR	$P = 2.75$ KW $Q = 92$ VAR	$P = 2.79$ KW $Q = 91$ VAR
			Type-2 fuzzy	$P = 2.8$ KW $Q = 90$ VAR	$P = 2.81$ KW $Q = 92$ VAR	$P = 2.78$ KW $Q = 91$ VAR	$P = 2.75$ KW $Q = 92$ VAR	$P = 2.79$ KW $Q = 91$ VAR

Table 4.44 Real and reactive powers of single-tuned SHAF with proposed Apq-SRF control under transient state.

Load	Reference controller	Without SAF	DC-link controller	With single-tuned SHAF (transient state)				
				PWM	HCC	AHCC	AFHCC type-1	AFHCC type-2
Diode bridge rectifier with RL load	Proposed Apq-SRF	$P = 3$ KW $Q = 1540$ VAR	PI	$P = 2.8$ KW $Q = 90$ VAR	$P = 2.81$ KW $Q = 92$ VAR	$P = 2.78$ KW $Q = 91$ VAR	$P = 2.7$ KW $Q = 92$ VAR	$P = 2.79$ KW $Q = 90$ VAR
			Type-1 fuzzy	$P = 2.8$ KW $Q = 90$ VAR	$P = 2.81$ KW $Q = 92$ VAR	$P = 2.78$ KW $Q = 91$ VAR	$P = 2.75$ KW $Q = 91$ VAR	$P = 2.79$ KW $Q = 91$ VAR
			Type-2 fuzzy	$P = 2.8$ KW $Q = 90$ VAR	$P = 2.8$ KW $Q = 92$ VAR	$P = 2.8$ KW $Q = 91$ VAR	$P = 2.75$ KW $Q = 92$ VAR	$P = 2.79$ KW $Q = 90$ VAR

Table 4.45 Real and reactive powers of double-tuned SHAF with proposed Apq-SRF control under transient state.

Load	Reference controller	Without SAF	DC-link controller	With SAF (transient state)				
				PWM	HCC	AHCC	AFHCC type-1	AFHCC type-2
Diode bridge rectifier with RL load	Proposed Apq-SRF	$P = 3$ KW $Q = 1540$ VAR	PI	$P = 2.8$ KW $Q = 90$ VAR	$P = 2.81$ KW $Q = 92$ VAR	$P = 2.78$ KW $Q = 91$ VAR	$P = 2.7$ KW $Q = 92$ VAR	$P = 2.79$ KW $Q = 91$ VAR
			Type-1 fuzzy	$P = 2.8$ KW $Q = 90$ VAR	$P = 2.81$ KW $Q = 92$ VAR	$P = 2.78$ KW $Q = 91$ VAR	$P = 2.75$ KW $Q = 91$ VAR	$P = 2.79$ KW $Q = 91$ VAR
			Type-2 fuzzy	$P = 2.8$ KW $Q = 90$ VAR	$P = 2.81$ KW $Q = 92$ VAR	$P = 2.78$ KW $Q = 91$ VAR	$P = 2.75$ KW $Q = 92$ VAR	$P = 2.79$ KW $Q = 91$ VAR

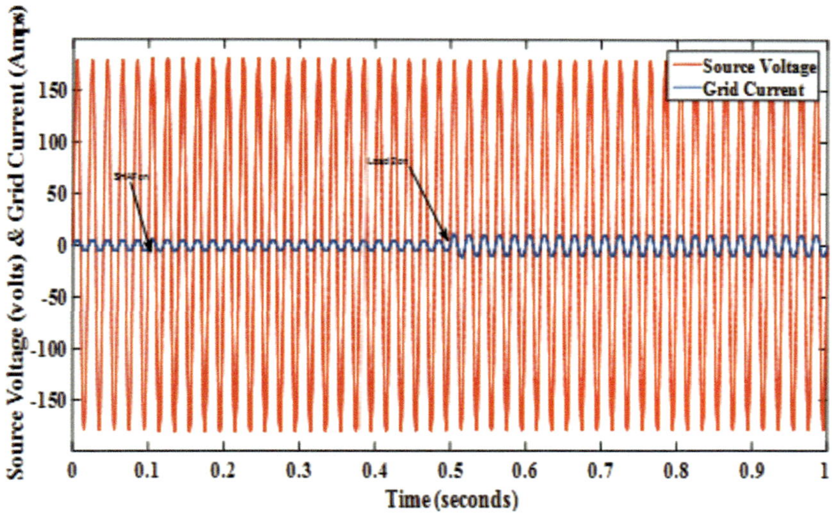

Figure 4.99 Power factor correction by the SHAF.

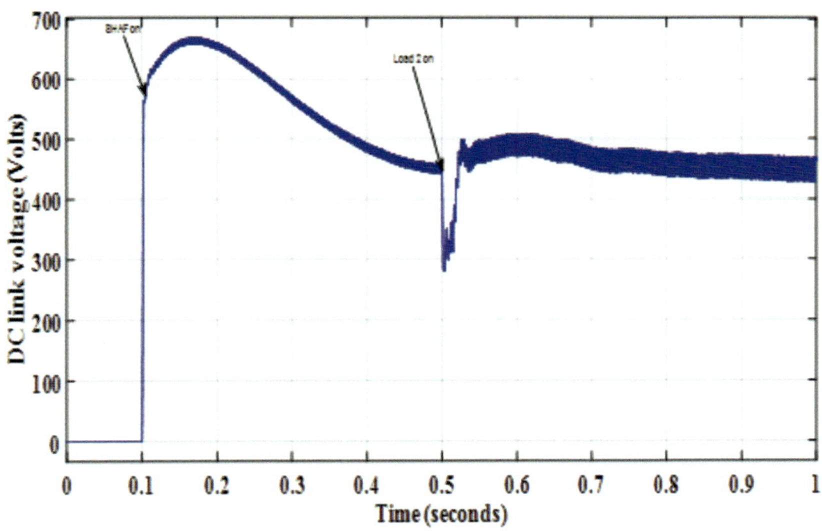

Figure 4.100 DC-link voltage across the VSI capacitor.

to maintain the constant DC-link voltage across VSI capacitor bank under steady- and transient-state conditions.

Figure 4.101 shows the reactive power balance, power factor correction, and real power demand under steady- and transient-state conditions using

SHAF. In Figure 4.101, the SHAF utilizes type-2 fuzzy logic DC-link voltage controller, AFHCC type-2 pulse generator, and double-tuned passive filter. In a similar manner, the oscillating component of reactive power, real power, and power factor has been eliminated by all the proposed SAF, single-, and double-tuned SHAF.

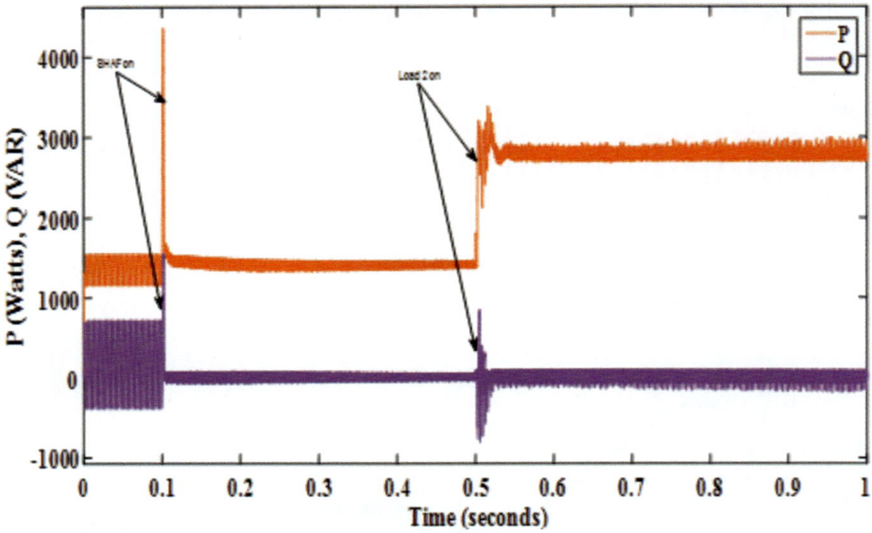

Figure 4.101 Real power, reactive power, and power factor correction.

Table 4.46 SHAF simulation parameters.

Parameters	**Values**
Supply voltage/frequency	220 volts/50 Hz
Source impedance (R_s, L_s)	0.1 Ω, 0.15 mH
Line impedance (R_{line}, L_{line})	0.4 Ω, 3.55 mH
Diode rectifier	6-diode
Load resistor (R_L) and inductor (L_L)	60 Ω, 20 mH
Interface inductor (L_c)	15 mH
DC-side capacitance (C_{dc})	35 μF
Reference voltage ($V_{dc,\ ref}$)	440 V
Voltage source inverter	6 IGBTS/Diodes
K_P and K_i values for PI controller	0.2, 1.5
Two single-tuned passive filters parameters (L_5, C_5, L_7, C_7)	13.5 mH, 30 μF, 6.89 mH, 30 μF
Double-tuned passive filter parameters (L_1, C_1, L_2, C_2)	4.56 mH, 60 μF, 5.56e-4 mH, 5.5e-4 F

Remarks: The Apq-SRF theory with various PWM pulse generators-based SHAF is simulated and compensation principle is evaluated. The AFHCC type-2 reduces the switching power losses and improves the SHAF performance than other control techniques. From these results, it is proved that the THD of the grid current follows harmonic standards IEEE 519-2014.

The simulation parameters for the SAF, two single-tuned passive filters, and one double-tuned passive filter are listed in Table 4.46.

4.9 Comparative Analysis

The comparison of various reference current extraction techniques in conjunction with various DC-link voltage controllers and pulse generators is divided into the following:

1. Comparison of various reference current extraction techniques with PI DC-link voltage controller and PWM pulse generator under steady state and transient state.
2. Comparison of various reference current extraction techniques with PI DC-link voltage controller and HCC pulse generator under steady state and transient state.
3. Comparison of various reference current extraction techniques with PI DC-link voltage controller and AHCC pulse generator under steady state and transient state.
4. Comparison of various reference current extraction techniques with PI DC-link voltage controller and type-1 AFHCC pulse generator under steady state and transient state.
5. Comparison of various reference current extraction techniques with PI DC-link voltage controller and type-2 AFHCC pulse generator under steady state and transient state.
6. Comparison of various reference current extraction techniques with type-1 fuzzy DC-link voltage controller and PWM pulse generator under steady state and transient state.
7. Comparison of various reference current extraction techniques with type-1 fuzzy I DC-link voltage controller and HCC pulse generator under steady state and transient state.
8. Comparison of various reference current extraction techniques with type-1 fuzzy DC-link voltage controller and AHCC pulse generator under steady state and transient state.

9. Comparison of various reference current extraction techniques with type-1 fuzzy DC-link voltage controller and type-1 AFHCC pulse generator under steady state and transient state.
10. Comparison of various reference current extraction techniques with type-2 fuzzy DC-link voltage controller and type-2 AFHCC pulse generator under steady state and transient state.
11. Comparison of various reference current extraction techniques with type-2 fuzzy DC-link voltage controller and PWM pulse generator under steady state and transient state.
12. Comparison of various reference current extraction techniques with type-2 fuzzy I DC-link voltage controller and HCC pulse generator under steady state and transient state.
13. Comparison of various reference current extraction techniques with type-2 fuzzy DC-link voltage controller and AHCC pulse generator under steady state and transient state.
14. Comparison of various reference current extraction techniques with type-2 fuzzy DC-link voltage controller and type-1 AFHCC pulse generator under steady state and transient state.
15. Comparison of various reference current extraction techniques with type-2 fuzzy DC-link voltage controller and type-2 AFHCC pulse generator under steady state and transient state.

4.9.1 Comparison of various reference current extraction techniques with PI DC-link voltage controller and PWM pulse generator

The performance comparison of various reference current extraction techniques with PI DC capacitor voltage regulator and PWM pulse generator for SAF is as shown in Figure 4.102. It has been observed that, all the reference current extraction techniques are able to cancel out grid current harmonics and excess reactive power as per the IEE-519-2022 standards. The SAF with proposed Apq-SRF reference current extraction technique is able to cancel out grid current harmonics and excess reactive power in conjunction with PI DC-link voltage controller and PWM pulse generator as shown in Figure 4.102.

The performance comparison of various reference current extraction techniques with PI DC capacitor voltage regulator and PWM pulse generator for single-tuned SHAF is as shown in Figure 4.103. It has been observed that, all the reference current extraction techniques are able to cancel out grid current harmonics and excess reactive power as per the IEE-519-2022 standards. The single-tuned SHAF with proposed Apq-SRF reference current extraction technique is able to cancel out grid current harmonics and excess

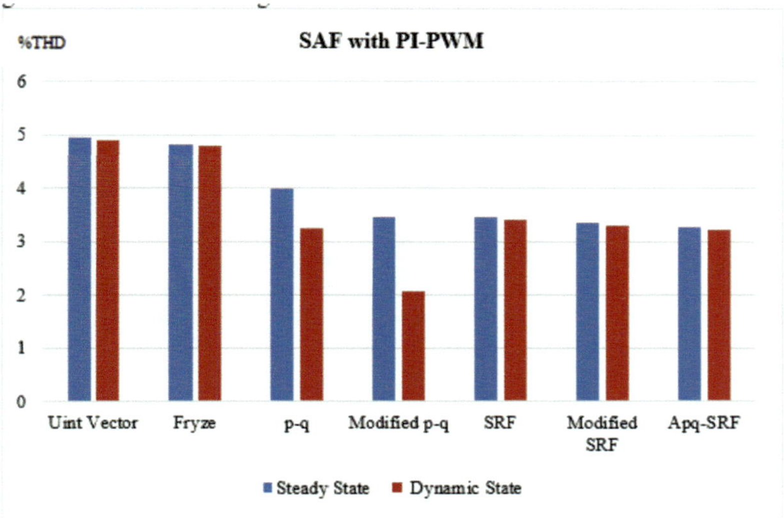

Figure 4.102 Comparison of SAF reference current extraction techniques with PI DC-link voltage controller and PWM pulse generator.

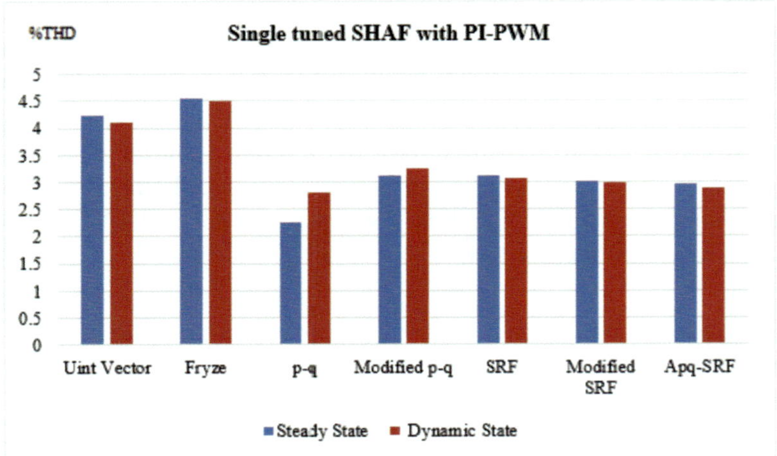

Figure 4.103 Comparison of single-tuned SHAF reference current extraction techniques with PI DC-link voltage controller and PWM pulse generator.

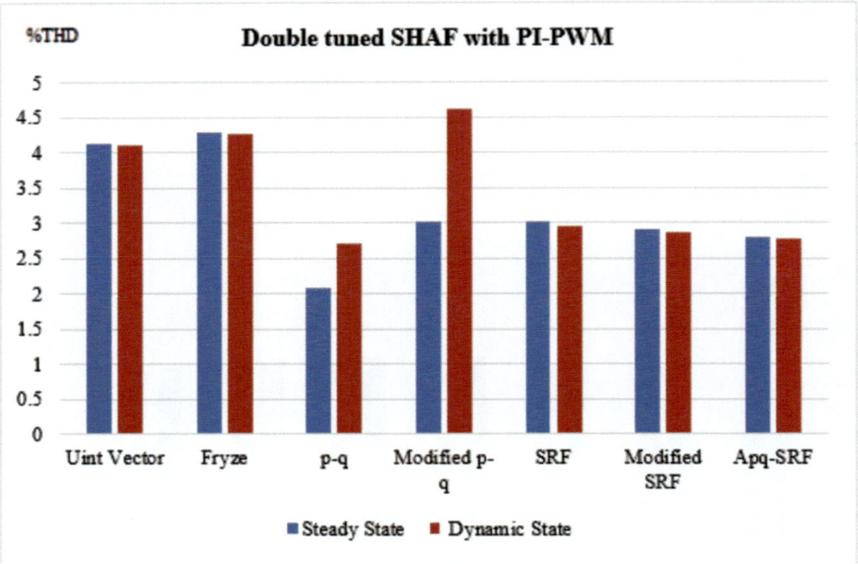

Figure 4.104 Comparison of double-tuned SHAF reference current extraction techniques with PI DC-link voltage controller and PWM pulse generator.

reactive power in conjunction with PI DC-link voltage controller and PWM pulse generator as shown in Figure 4.103.

The performance comparison of various reference current extraction techniques with PI DC capacitor voltage regulator and PWM pulse generator for double-tuned SHAF is as shown in Figure 4.104. It has been observed that, all the reference current extraction techniques are able to cancel out grid current harmonics and excess reactive power as per the IEE-519-2022 standards. The double-tuned SHAF with proposed Apq-SRF reference current extraction technique is able to cancel out grid current harmonics and excess reactive power in conjunction with PI DC-link voltage controller and PWM pulse generator as shown in Figure 4.104.

4.9.2 Comparison of various reference current extraction techniques with PI DC-link voltage controller and HCC pulse generator

The performance comparison of various reference current extraction techniques with PI DC capacitor voltage regulator and HCC pulse generator for SAF is as shown in Figure 4.105. It has been observed that, all the reference

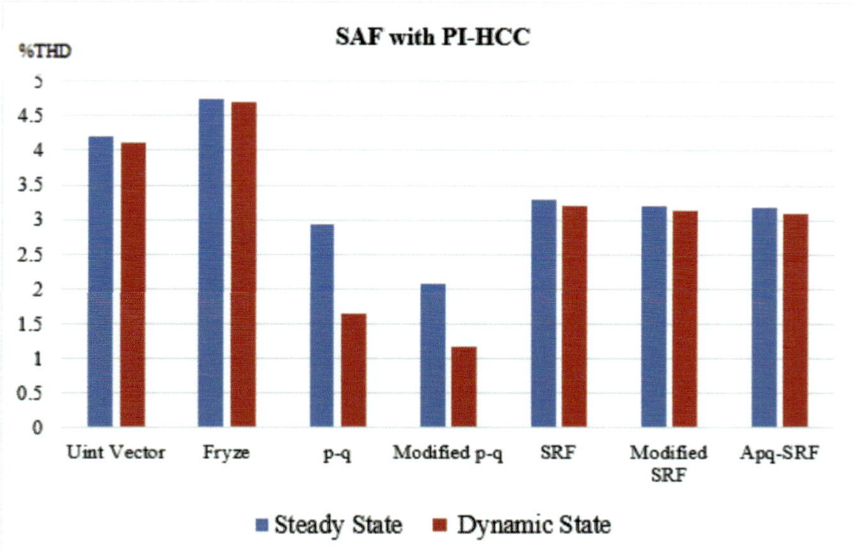

Figure 4.105 Comparison of SAF reference current extraction techniques with PI DC-link voltage controller and HCC pulse generator.

current extraction techniques are able to cancel out grid current harmonics and excess reactive power as per the IEE-519-2022 standards. The SAF with proposed Apq-SRF reference current extraction technique is able to cancel out grid current harmonics and excess reactive power in conjunction with PI DC-link voltage controller and HCC pulse generator as shown in Figure 4.105.

The performance comparison of various reference current extraction techniques with PI DC capacitor voltage regulator and HCC pulse generator for single-tuned SHAF is as shown in Figure 4.104. It has been observed that, all the reference current extraction techniques are able to cancel out grid current harmonics and excess reactive power as per the IEE-519-2022 standards. The single-tuned SHAF with proposed Apq-SRF reference current extraction technique is able to cancel out grid current harmonics and excess reactive power in conjunction with PI DC-link voltage controller and HCC pulse generator as shown in Figure 4.106.

The performance comparison of various reference current extraction techniques with PI DC capacitor voltage regulator and HCC pulse generator for double-tuned SHAF is as shown in Figure 4.107. It has been observed that, all

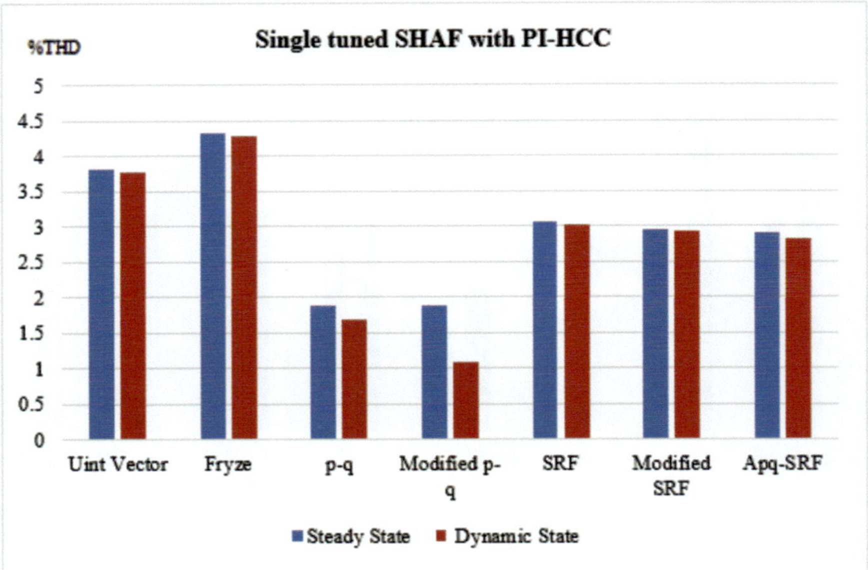

Figure 4.106 Comparison of single-tuned SHAF reference current extraction techniques with PI DC-link voltage controller and HCC pulse generator.

the reference current extraction techniques are able to cancel out grid current harmonics and excess reactive power as per the IEE-519-2022 standards. The double-tuned SHAF with proposed Apq-SRF reference current extraction technique is able to cancel out grid current harmonics and excess reactive power in conjunction with PI DC-link voltage controller and HCC pulse generator as shown in Figure 4.107.

4.9.3 Comparison of various reference current control techniques with PI DC-link voltage controller and AHCC pulse generator

The performance comparison of various reference current extraction techniques with PI DC capacitor voltage regulator and AHCC pulse generator for SAF is as shown in Figure 4.108. It has been observed that, all the reference current extraction techniques are able to cancel out grid current harmonics and excess reactive power as per the IEE-519-2022 standards. The SAF with proposed Apq-SRF reference current extraction technique is able to cancel out grid current harmonics and excess reactive power in conjunction

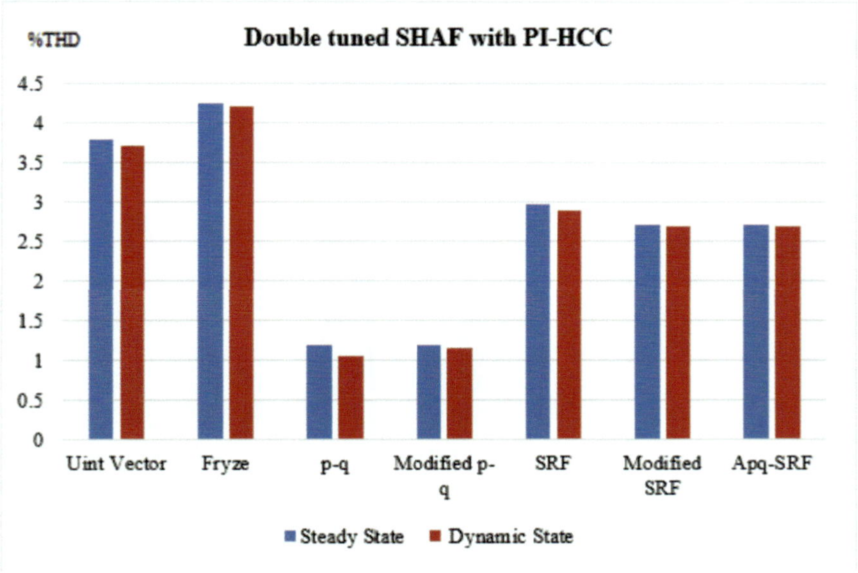

Figure 4.107 Comparison of double-tuned SHAF reference current extraction techniques with PI DC-link voltage controller and HCC pulse generator.

with PI DC-link voltage controller and AHCC pulse generator as shown in Figure 4.108.

The performance comparison of various reference current extraction techniques with PI DC capacitor voltage regulator and AHCC pulse generator for single-tuned SHAF is as shown in Figure 4.109. It has been observed that, all the reference current extraction techniques are able to cancel out grid current harmonics and excess reactive power as per the IEE-519-2022 standards. The single-tuned SHAF with proposed Apq-SRF reference current extraction technique is able to cancel out grid current harmonics and excess reactive power in conjunction with PI DC-link voltage controller and AHCC pulse generator as shown in Figure 4.109.

The performance comparison of various reference current extraction techniques with PI DC capacitor voltage regulator and AHCC pulse generator for double-tuned SHAF is as shown in Figure 4.110. It has been observed that, all the reference current extraction techniques are able to cancel out grid current harmonics and excess reactive power as per the IEE-519-2022 standards. The double-tuned SHAF with proposed Apq-SRF reference current extraction technique is able to cancel out grid current harmonics and excess reactive

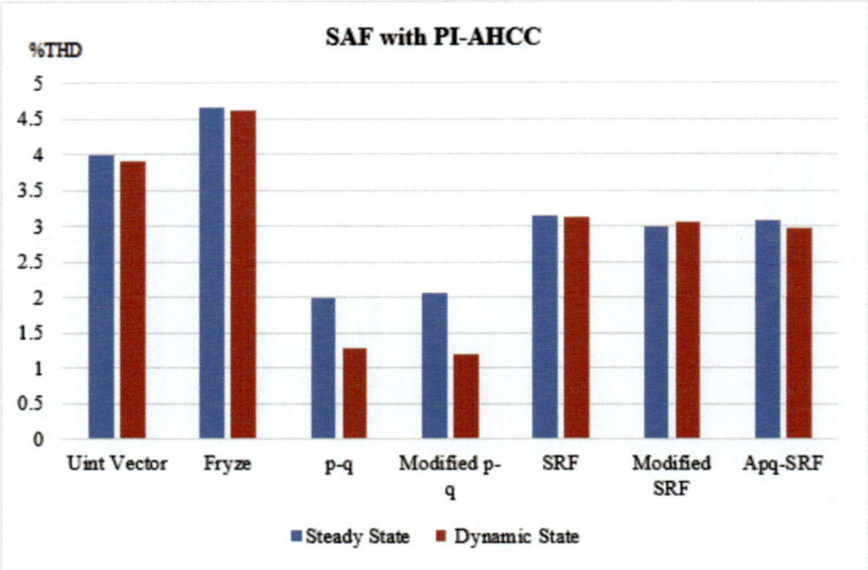

Figure 4.108 Comparison of SAF reference current extraction techniques with PI DC-link voltage controller and AHCC pulse generator.

power in conjunction with PI DC-link voltage controller and AHCC pulse generator as shown in Figure 4.110.

4.9.4 Comparison of various reference current extraction techniques with PI DC-link voltage controller and type-1 AFHCC pulse generator

The performance comparison of various reference current extraction techniques with PI DC capacitor voltage regulator and type-1 AFHCC pulse generator for SAF is as shown in Figure 4.111. It has been observed that, all the reference current extraction techniques are able to cancel out grid current harmonics and excess reactive power as per the IEE-519-2022 standards. The SAF with proposed Apq-SRF reference current extraction technique is able to cancel out grid current harmonics and excess reactive power in conjunction with PI DC-link voltage controller and type-1 AFHCC pulse generator as shown in Figure 4.111.

The performance comparison of various reference current extraction techniques with PI DC capacitor voltage regulator and type-1 AFHCC pulse generator for single-tuned SHAF is as shown in Figure 4.112. It has been

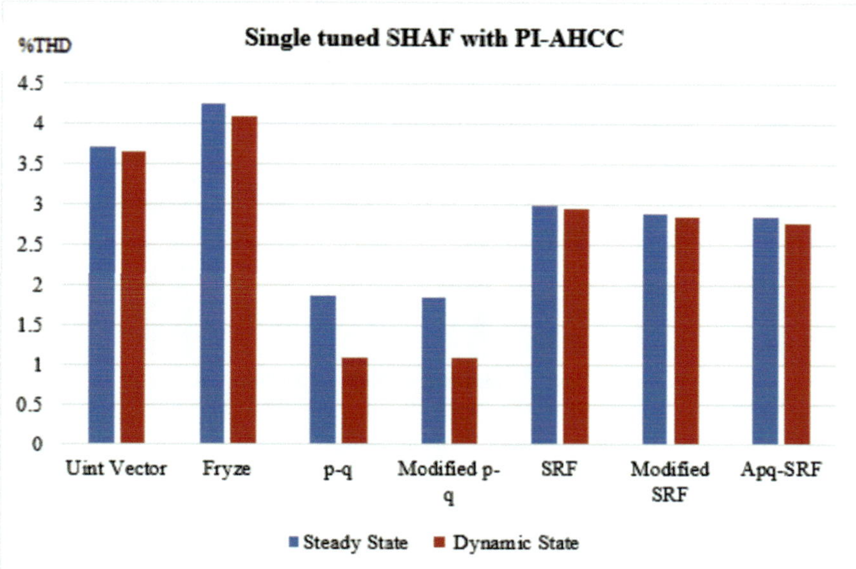

Figure 4.109 Comparison of single-tuned SHAF reference current extraction techniques with PI DC-link voltage controller and AHCC pulse generator.

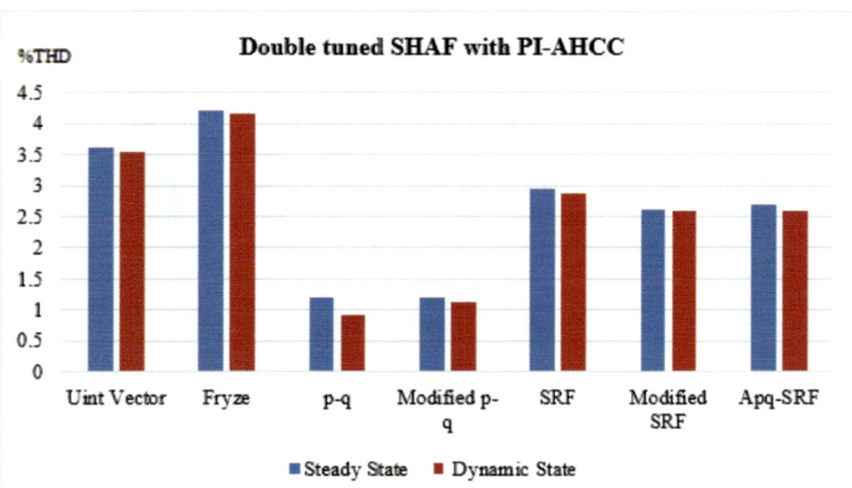

Figure 4.110 Comparison of double-tuned SHAF reference current extraction techniques with PI DC-link voltage controller and AHCC pulse generator.

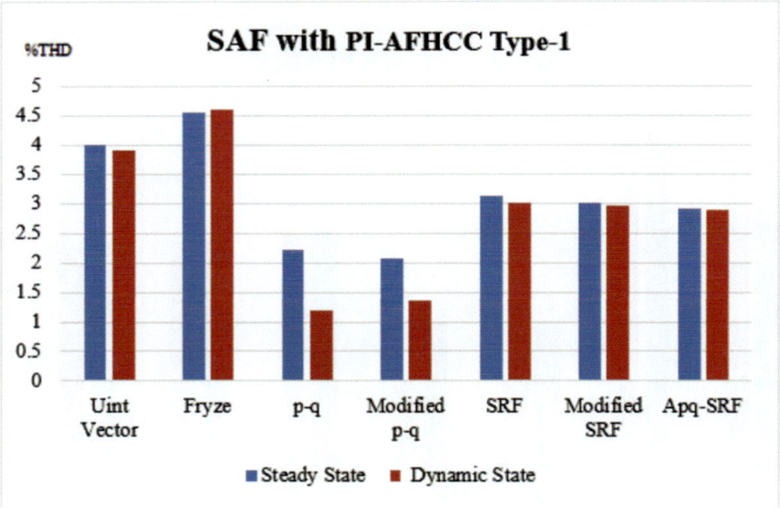

Figure 4.111 Comparison of SAF reference current extraction techniques with PI DC-link voltage controller and type-1 AFHCC pulse generator.

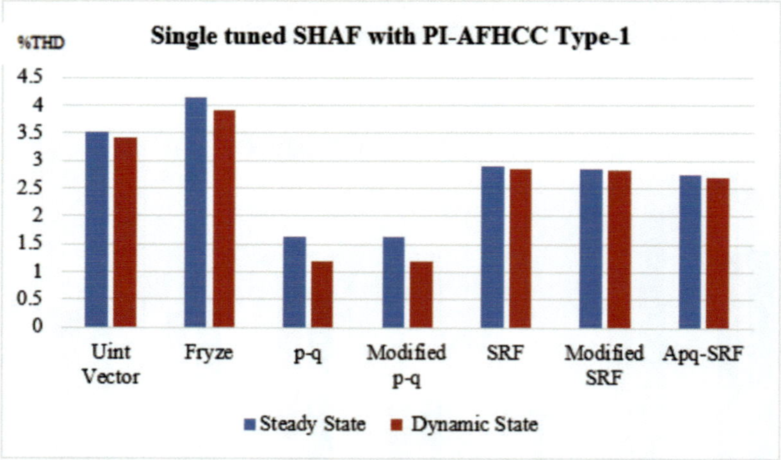

Figure 4.112 Comparison of single-tuned SHAF reference current extraction techniques with PI DC-link voltage controller and type-1 AFHCC pulse generator.

observed that, all the reference current extraction techniques are able to cancel out grid current harmonics and excess reactive power as per the IEE-519-2022 standards. The single-tuned SHAF with proposed Apq-SRF

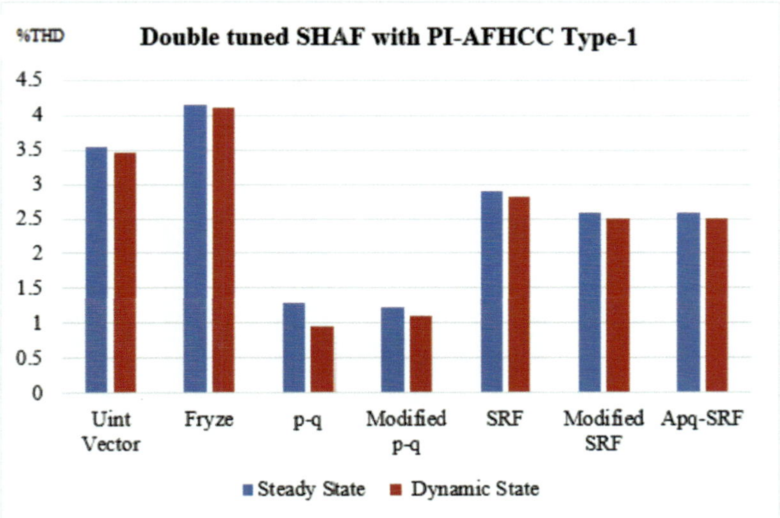

Figure 4.113 Comparison of double-tuned SHAF reference current extraction techniques with PI DC-link voltage controller and type-1 AFHCC pulse generator.

reference current extraction technique is able to cancel out grid current harmonics and excess reactive power in conjunction with PI DC-link voltage controller and type-1 AFHCC pulse generator as shown in Figure 4.112.

The performance comparison of various reference current extraction techniques with PI DC capacitor voltage regulator and type-1 AFHCC pulse generator for double-tuned SHAF is as shown in Figure 4.113. It has been observed that, all the reference current extraction techniques are able to cancel out grid current harmonics and excess reactive power as per the IEE-519-2022 standards. The double-tuned SHAF with proposed Apq-SRF reference current extraction technique is able to cancel out grid current harmonics and excess reactive power in conjunction with PI DC-link voltage controller and type-1 AFHCC pulse generator as shown in Figure 4.113.

4.9.5 Comparison of various reference current extraction techniques with PI DC-link voltage controller and type-2 AFHCC pulse generator

The performance comparison of various reference current extraction techniques with PI DC capacitor voltage regulator and type-2 AFHCC pulse generator for SAF is as shown in Figure 4.114. It has been observed that, all

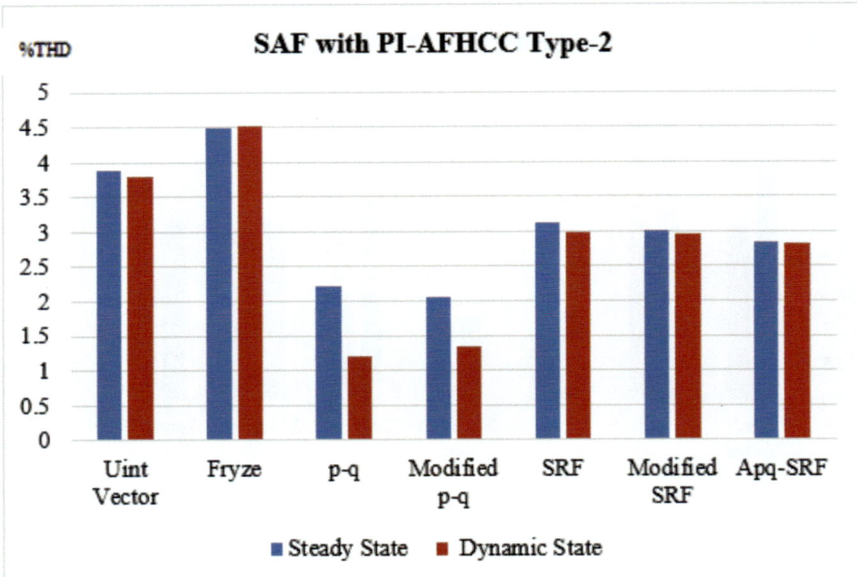

Figure 4.114 Comparison of SAF reference current extraction techniques with PI DC-link voltage controller and type-2 AFHCC pulse generator.

the reference current extraction techniques are able to cancel out grid current harmonics and excess reactive power as per the IEE-519-2022 standards. The SAF with proposed Apq-SRF reference current extraction technique is able to cancel out grid current harmonics and excess reactive power in conjunction with PI DC-link voltage controller and type-2 AFHCC pulse generator as shown in Figure 4.114.

The performance comparison of various reference current extraction techniques with PI DC capacitor voltage regulator and type-2 AFHCC pulse generator for single-tuned SHAF is as shown in Figure 4.115. It has been observed that, all the reference current extraction techniques are able to cancel out grid current harmonics and excess reactive power as per the IEE-519-2022 standards. The single-tuned SHAF with proposed Apq-SRF reference current extraction technique is able to cancel out grid current harmonics and excess reactive power in conjunction with PI DC-link voltage controller and type-2 AFHCC pulse generator as shown in Figure 4.115.

The performance comparison of various reference current extraction techniques with PI DC capacitor voltage regulator and type-2 AFHCC pulse generator for double-tuned SHAF is as shown in Figure 4.116. It has been

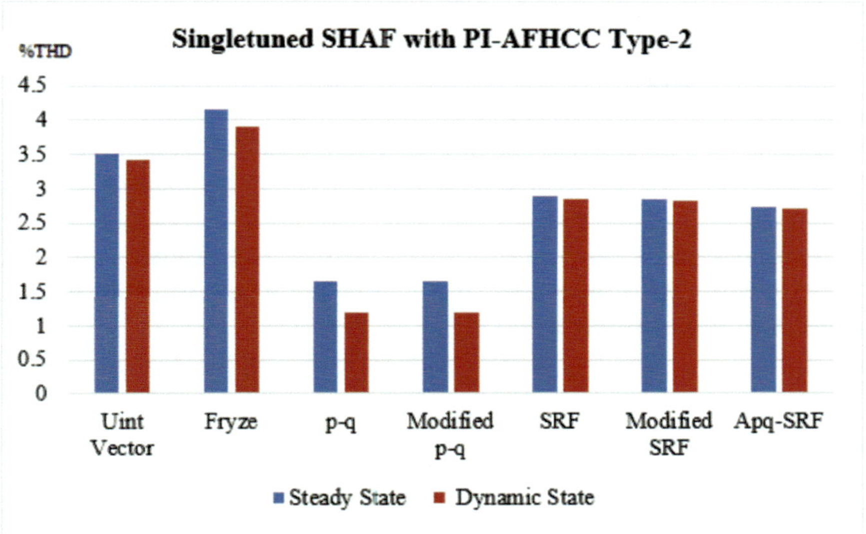

Figure 4.115 Comparison of single-tuned SHAF reference current extraction techniques with PI DC-link voltage controller and type-2 AFHCC pulse generator.

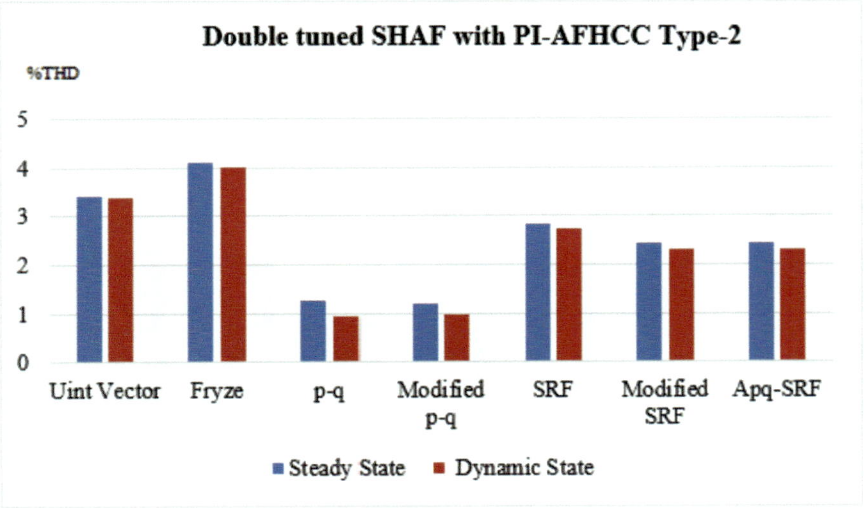

Figure 4.116 Comparison of double-tuned SHAF reference current extraction techniques with PI DC-link voltage controller and type-2 AFHCC pulse generator.

observed that, all the reference current extraction techniques are able to cancel out grid current harmonics and excess reactive power as per the IEE-519-2022 standards. The double-tuned SHAF with proposed Apq-SRF reference current extraction technique is able to cancel out grid current harmonics and excess reactive power in conjunction with PI DC- link voltage controller and type-2 AFHCC pulse generator as shown in Figure 4.116.

4.9.6 Comparison of various reference current extraction techniques with type-1 fuzzy DC-link voltage controller and PWM pulse generator

The performance comparison of various reference current extraction techniques with type-1 fuzzy DC capacitor voltage regulator and PWM pulse generator for SAF is as shown in Figure 4.117. It has been observed that, all the reference current extraction techniques are able to cancel out grid current harmonics and excess reactive power as per the IEE-519-2022 standards. The SAF with proposed Apq-SRF reference current extraction technique is able to cancel out grid current harmonics and excess reactive power in conjunction with type-1 fuzzy DC capacitor voltage controller and PWM pulse generator as shown in Figure 4.117.

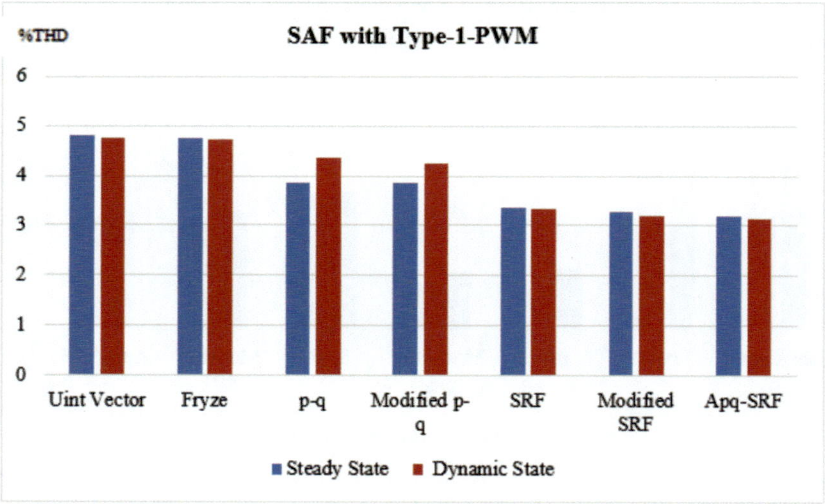

Figure 4.117 Comparison of SAF reference current extraction techniques with type-1 fuzzy DC-link voltage controller and PWM pulse generator.

The performance comparison of various reference current extraction techniques with type-1 fuzzy DC capacitor voltage regulator and PWM pulse generator for single-tuned SHAF is as shown in Figure 4.118. It has been observed that, all the reference current extraction techniques are able to cancel out grid current harmonics and excess reactive power as per the IEE-519-2022 standards. The single-tuned SHAF with proposed Apq-SRF reference current extraction technique is able to cancel out grid current harmonics and excess reactive power in conjunction with type-1 fuzzy DC capacitor voltage controller and PWM pulse generator as shown in Figure 4.118.

The performance comparison of various reference current extraction techniques with type-1 fuzzy DC capacitor voltage regulator and PWM pulse generator for double-tuned SHAF is as shown in Figure 4.119. It has been observed that, all the reference current extraction techniques are able to cancel out grid current harmonics and excess reactive power as per the IEE-519-2022 standards. The double-tuned SHAF with proposed Apq-SRF reference current extraction technique is able to cancel out grid current harmonics and excess reactive power in conjunction with type-1 fuzzy DC capacitor voltage controller and PWM pulse generator as shown in Figure 4.119.

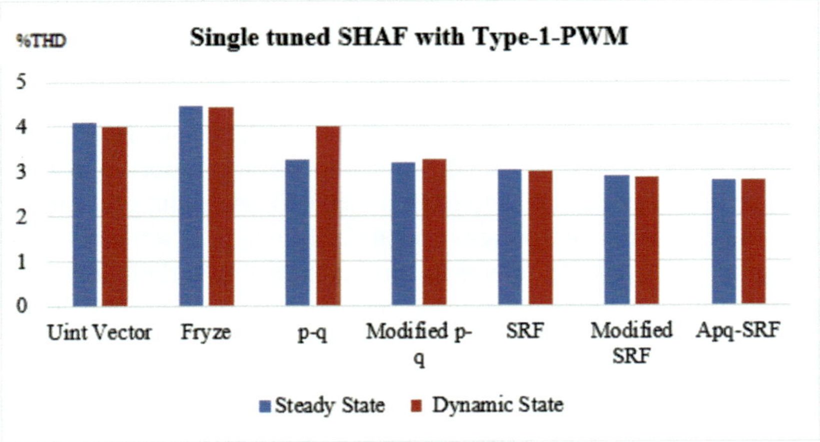

Figure 4.118 Comparison of single-tuned SHAF reference current extraction techniques with type-1 fuzzy DC-link voltage controller and PWM pulse generator.

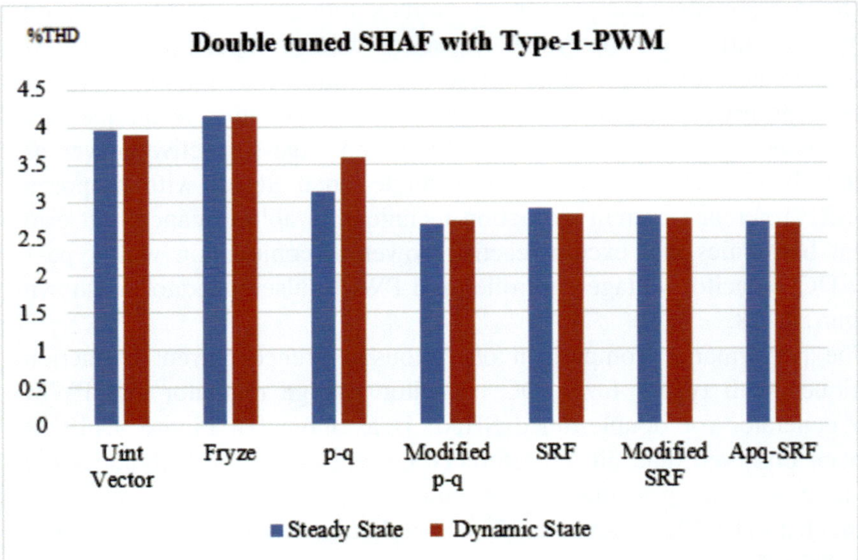

Figure 4.119 Comparison of double-tuned SHAF reference current extraction techniques with type-1 fuzzy DC-link voltage controller and PWM pulse generator.

4.9.7 Comparison of various reference current extraction techniques with type-1 fuzzy DC-link voltage controller and HCC pulse generator

The performance comparison of various reference current extraction techniques with type-1 fuzzy DC capacitor voltage regulator and HCC pulse generator for SAF is as shown in Figure 4.120. It has been observed that, all the reference current extraction techniques are able to cancel out grid current harmonics and excess reactive power as per the IEE-519-2022 standards. The SAF with proposed Apq-SRF reference current extraction technique is able to cancel out grid current harmonics and excess reactive power in conjunction with type-1 fuzzy DC capacitor voltage controller and HCC pulse generator as shown in Figure 4.120.

The performance comparison of various reference current extraction techniques with type-1 fuzzy DC capacitor voltage regulator and HCC pulse generator for single-tuned SHAF is as shown in Figure 4.121. It has been observed that, all the reference current extraction techniques are able to cancel out grid current harmonics and excess reactive power as per the IEE-519-2022 standards. The single-tuned SHAF with proposed

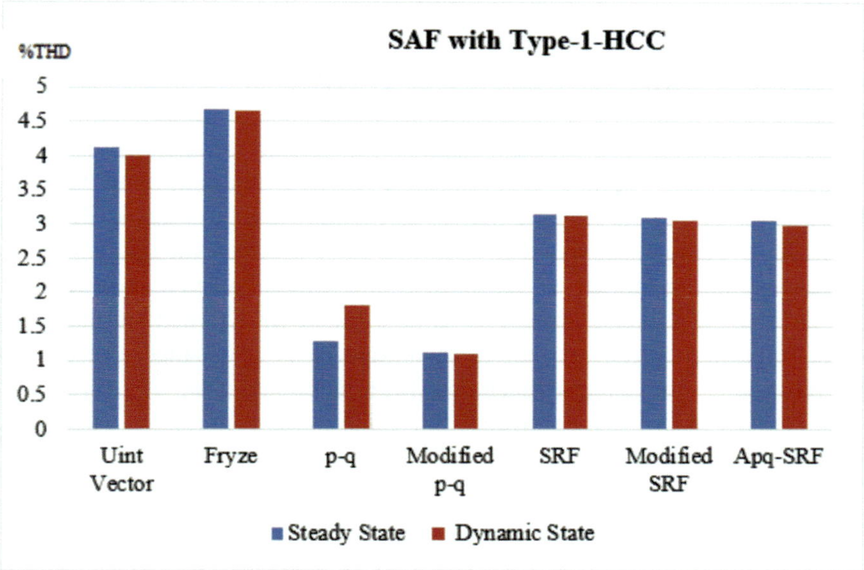

Figure 4.120 Comparison of SAF reference current extraction techniques with type-1 fuzzy DC-link voltage controller and HCC pulse generator.

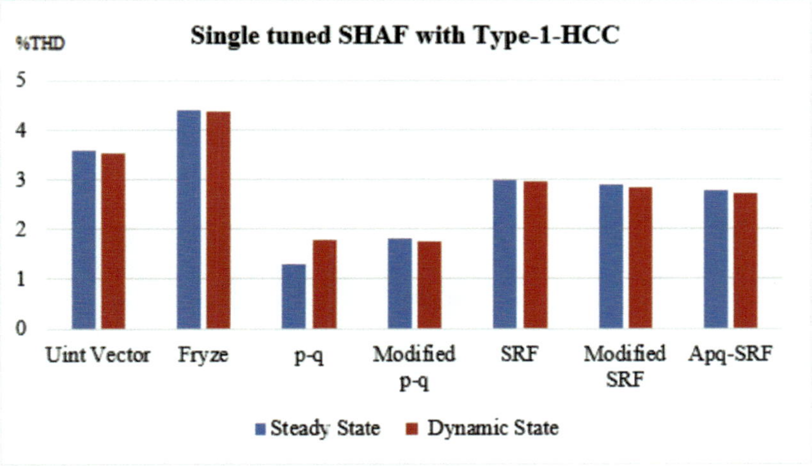

Figure 4.121 Comparison of single-tuned SHAF reference current extraction techniques with type-1 fuzzy DC-link voltage controller and HCC pulse generator.

Apq-SRF reference current extraction technique is able to cancel out grid current harmonics and excess reactive power in conjunction with type-1 fuzzy DC capacitor voltage controller and HCC pulse generator as shown in Figure 4.121.

The performance comparison of various reference current extraction techniques with type-1 fuzzy DC capacitor voltage regulator and HCC pulse generator for double-tuned SHAF is as shown in Figure 4.122. It has been observed that, all the reference current extraction techniques are able to cancel out grid current harmonics and excess reactive power as per the IEE-519-2022 standards. The double-tuned SHAF with proposed Apq-SRF reference current extraction technique is able to cancel out grid current harmonics and excess reactive power in conjunction with type-1 fuzzy DC capacitor voltage controller and HCC pulse generator as shown in Figure 4.122.

4.9.8 Comparison of various reference current extraction techniques with type-1 fuzzy DC-link voltage controller and AHCC pulse generator

The performance comparison of various reference current extraction techniques with type-1 fuzzy DC capacitor voltage regulator and AHCC pulse generator for SAF is as shown in Figure 4.123. It has been observed that, all

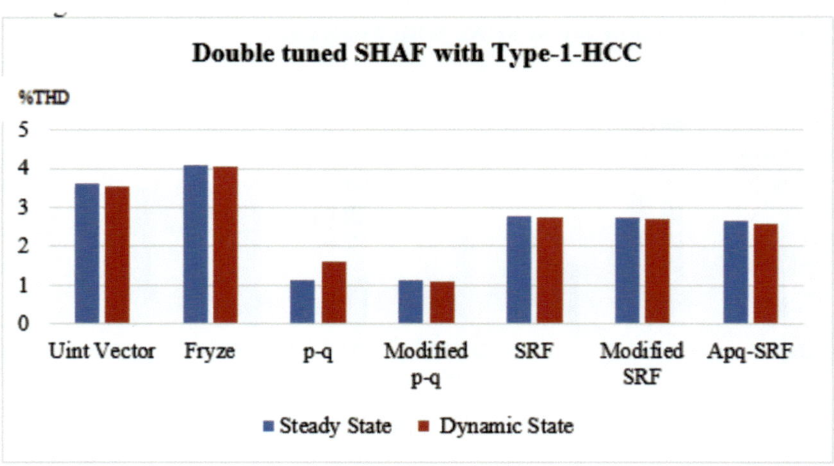

Figure 4.122 Comparison of double-tuned SHAF reference current extraction techniques with type-1 fuzzy DC-link voltage controller and HCC pulse generator.

the reference current extraction techniques are able to cancel out grid current harmonics and excess reactive power as per the IEE-519-2022 standards. The SAF with proposed Apq-SRF reference current extraction technique is able to cancel out grid current harmonics and excess reactive power in conjunction with type-1 fuzzy DC capacitor voltage controller and AHCC pulse generator as shown in Figure 4.123.

The performance comparison of various reference current extraction techniques with type-1 fuzzy DC capacitor voltage regulator and AHCC pulse generator for single-tuned SHAF is as shown in Figure 4.124. It has been observed that, all the reference current extraction techniques are able to cancel out grid current harmonics and excess reactive power as per the IEE-519-2022 standards. The single-tuned SHAF with proposed Apq-SRF reference current extraction technique is able to cancel out grid current harmonics and excess reactive power in conjunction with type-1 fuzzy DC capacitor voltage controller and AHCC pulse generator as shown in Figure 4.124.

The performance comparison of various reference current extraction techniques with type-1 fuzzy DC capacitor voltage regulator and AHCC pulse generator for double-tuned SHAF is as shown in Figure 4.125. It has been observed that, all the reference current extraction techniques are able to cancel out grid current harmonics and excess reactive power as per the IEE-519-2022 standards. The double-tuned SHAF with proposed

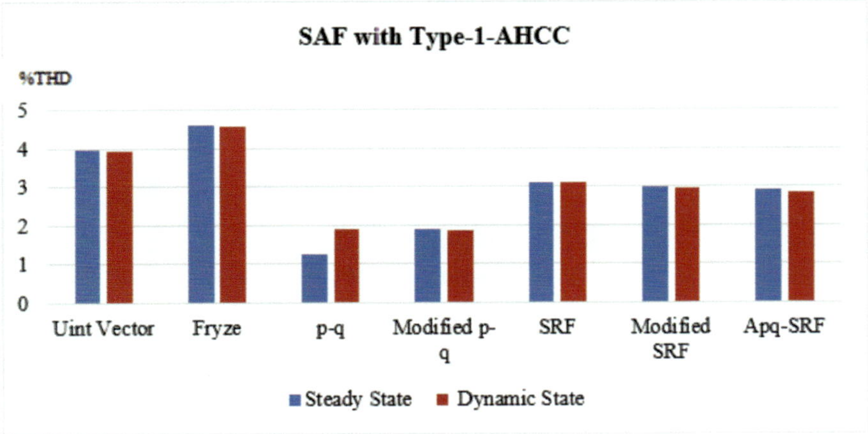

Figure 4.123 Comparison of SAF reference current extraction techniques with type-1 fuzzy DC-link voltage controller and AHCC pulse generator.

Apq-SRF reference current extraction technique is able to cancel out grid current harmonics and excess reactive power in conjunction with type-1 fuzzy DC capacitor voltage controller and AHCC pulse generator as shown in Figure 4.125.

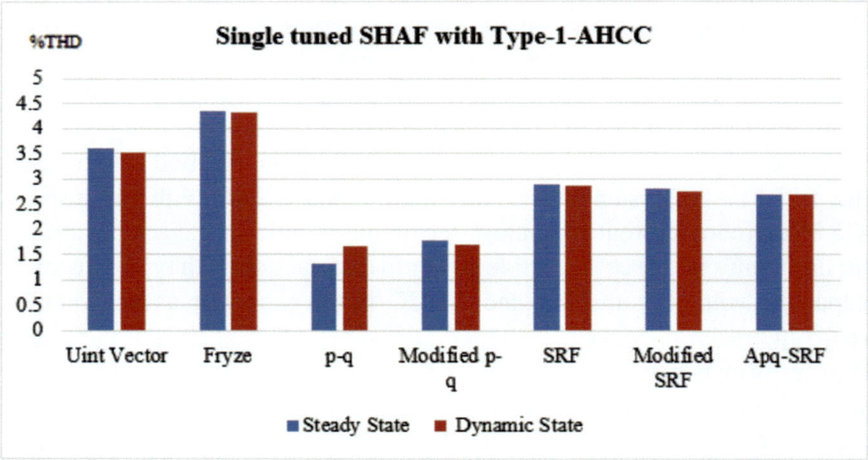

Figure 4.124 Comparison of single-tuned SHAF reference current extraction techniques with type-1 fuzzy DC-link voltage controller and AHCC pulse generator.

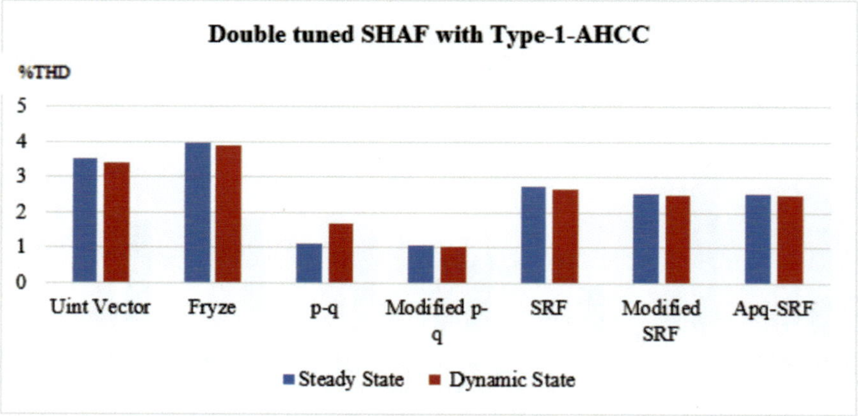

Figure 4.125 Comparison of double-tuned SHAF reference current extraction techniques with type-1 fuzzy DC-link voltage controller and AHCC pulse generator.

4.9.9 Comparison of various reference current extraction techniques with type-1 fuzzy DC-link voltage controller and type-1 AFHCC pulse generator

The performance comparison of various reference current extraction techniques with type-1 fuzzy DC capacitor voltage regulator and type-1 AFHCC pulse generator for SAF is as shown in Figure 4.126. It has been observed that, all the reference current extraction techniques are able to cancel out grid current harmonics and excess reactive power as per the IEE-519-2022 standards. The SAF with proposed Apq-SRF reference current extraction technique is able to cancel out grid current harmonics and excess reactive power in conjunction with type-1 fuzzy DC capacitor voltage controller and type-1 AFHCC as shown in Figure 4.126.

The performance comparison of various reference current extraction techniques with type-1 fuzzy DC capacitor voltage regulator and type-1 AFHCC pulse generator for single-tuned SHAF is as shown in Figure 4.127. It has been observed that, all the reference current extraction techniques are able to cancel out grid current harmonics and excess reactive power as per the IEE-519-2022 standards. The single-tuned SHAF with proposed Apq-SRF reference current extraction technique is able to cancel out grid current

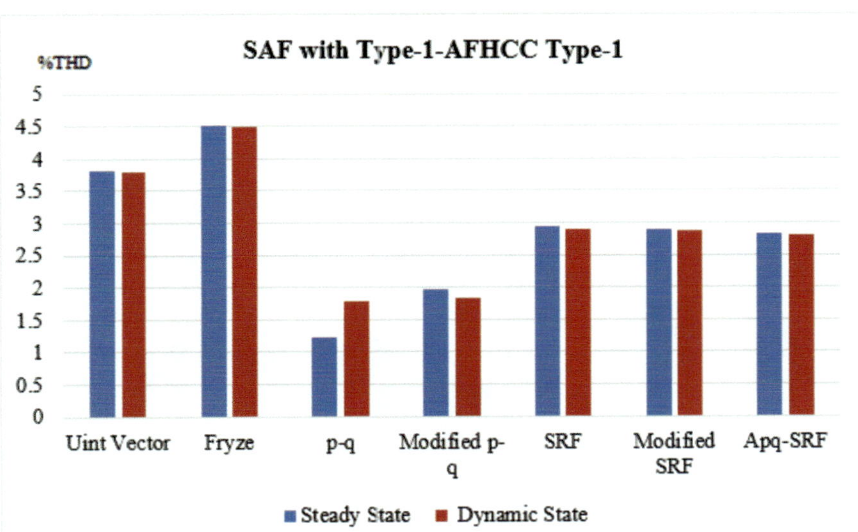

Figure 4.126 Comparison of SAF reference current extraction techniques with Type-1 fuzzy DC-link voltage controller and Type-1 AFHCC pulse generator.

harmonics and excess reactive power in conjunction with type-1 fuzzy DC capacitor voltage controller and type-1 AFHCC as shown in Figure 4.127.

The performance comparison of various reference current extraction techniques with type-1 fuzzy DC capacitor voltage regulator and type-1 AFHCC pulse generator for double-tuned SHAF is as shown in Figure 4.128. It has been observed that, all the reference current extraction techniques are able to cancel out grid current harmonics and excess reactive power as per the IEE-519-2022 standards. The double-tuned SHAF with proposed Apq-SRF reference current extraction technique is able to cancel out grid current

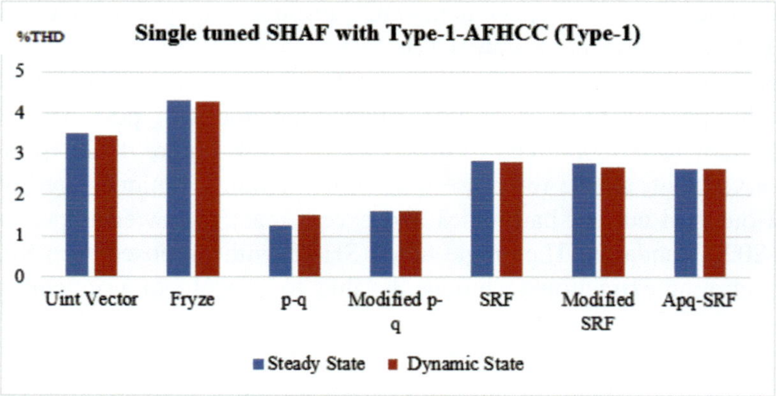

Figure 4.127 Comparison of single-tuned SHAF reference current extraction techniques with type-1 fuzzy DC-link voltage controller and type-1 AFHCC pulse generator.

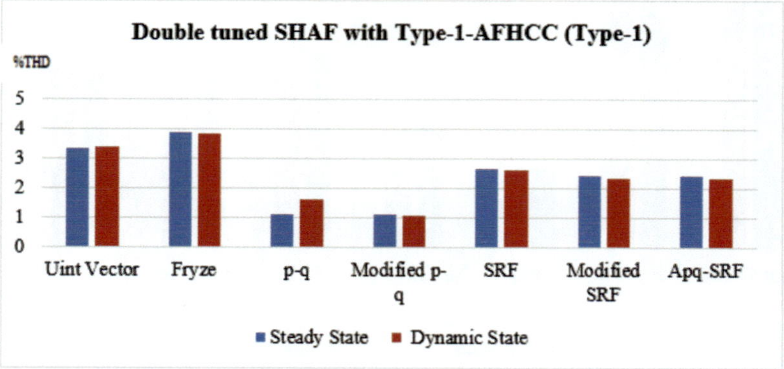

Figure 4.128 Comparison of double-tuned SHAF reference current extraction techniques with type-1 fuzzy DC-link voltage controller and type-1 AFHCC pulse generator.

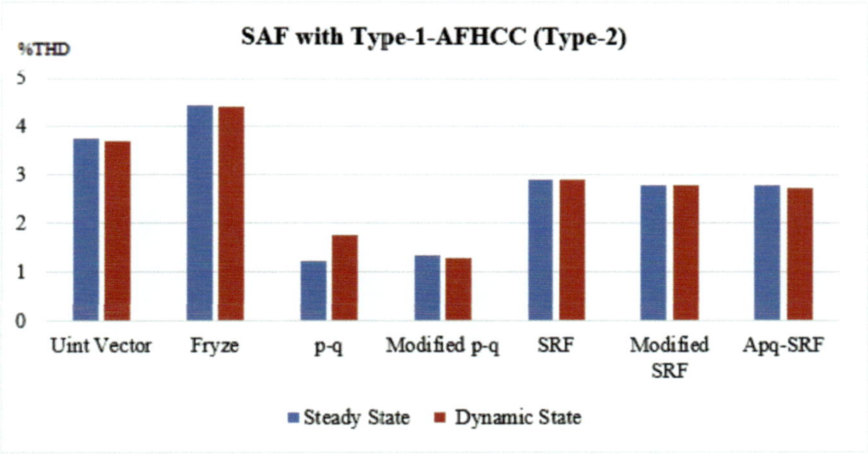

Figure 4.129 Comparison of SAF reference current extraction techniques with type-1 fuzzy DC-link voltage controller and type-2 AFHCC pulse generator.

harmonics and excess reactive power in conjunction with type-1 fuzzy DC capacitor voltage controller and type-1 AFHCC as shown in Figure 4.128.

4.9.10 Comparison of various reference current extraction techniques with type-1 fuzzy DC-link voltage controller and type-2 AFHCC pulse generator

The performance comparison of various reference current extraction techniques with type-1 fuzzy DC capacitor voltage regulator and type-2 AFHCC pulse generator for SAF is as shown in Figure 4.129. It has been observed that, all the reference current extraction techniques are able to cancel out grid current harmonics and excess reactive power as per the IEE-519-2022 standards. The SAF with proposed Apq-SRF reference current extraction technique is able to cancel out grid current harmonics and excess reactive power in conjunction with type-1 fuzzy DC capacitor voltage controller and type-2 AFHCC as shown in Figure 4.129.

The performance comparison of various reference current extraction techniques with type-1 fuzzy DC capacitor voltage regulator and type-2 AFHCC pulse generator for single-tuned SHAF is as shown in Figure 4.130. It has been observed that, all the reference current extraction techniques are able to cancel out grid current harmonics and excess reactive power as per the IEE-519-2022 standards. The single-tuned SHAF with proposed Apq-SRF

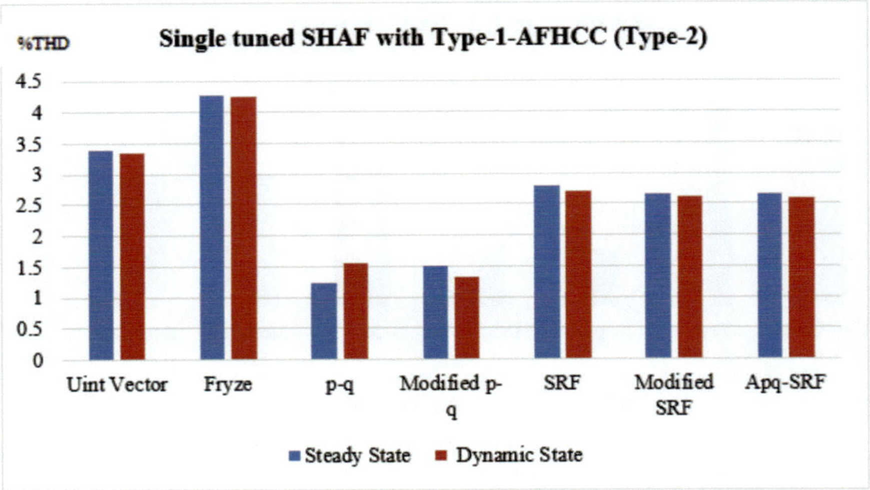

Figure 4.130 Comparison of single-tuned SHAF reference current extraction techniques with type-1 fuzzy DC-link voltage controller and type-2 AFHCC pulse generator.

reference current extraction technique is able to cancel out grid current harmonics and excess reactive power in conjunction with type-1 fuzzy DC capacitor voltage controller and type-2 AFHCC as shown in Figure 4.130.

The performance comparison of various reference current extraction techniques with type-1 fuzzy DC capacitor voltage regulator and type-2 AFHCC pulse generator for double-tuned SHAF is as shown in Figure 4.131. It has been observed that, all the reference current extraction techniques are able to cancel out grid current harmonics and excess reactive power as per the IEE-519-2022 standards. The double-tuned SHAF with proposed Apq-SRF reference current extraction technique is able to cancel out grid current harmonics and excess reactive power in conjunction with type-1 fuzzy DC capacitor voltage controller and type-2 AFHCC as shown in Figure 4.131.

4.9.11 Comparison of various reference current extraction techniques with type-2 fuzzy DC-link voltage controller and PWM pulse generator

The performance comparison of various reference current extraction techniques with type-2 fuzzy DC capacitor voltage regulator and PWM pulse generator for SAF is as shown in Figure 4.132. It has been observed that, all the reference current extraction techniques are able to cancel out grid current

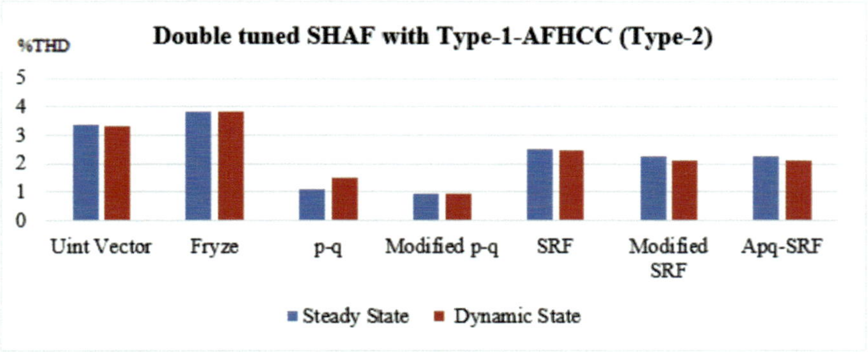

Figure 4.131 Comparison of single-tuned SHAF reference current extraction techniques with type-1 fuzzy DC-link voltage controller and type-2 AFHCC pulse generator.

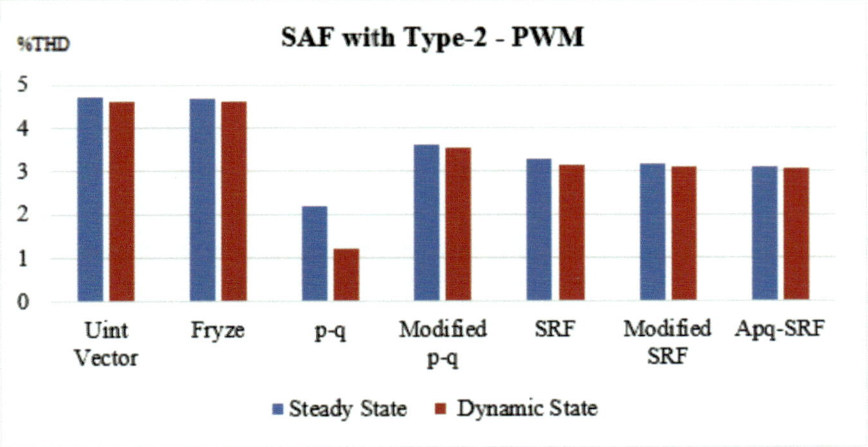

Figure 4.132 Comparison of SAF reference current extraction techniques with type-2 fuzzy DC-link voltage controller and PWM pulse generator.

harmonics and excess reactive power as per the IEE-519-2022 standards. The SAF with proposed Apq-SRF reference current extraction technique is able to cancel out grid current harmonics and excess reactive power in conjunction with type-2 fuzzy DC capacitor voltage controller and PWM pulse generator is as shown in Figure 4.132.

The performance comparison of various reference current extraction techniques with type-2 fuzzy DC capacitor voltage regulator and PWM pulse generator for single-tuned SHAF is as shown in Figure 4.133. It

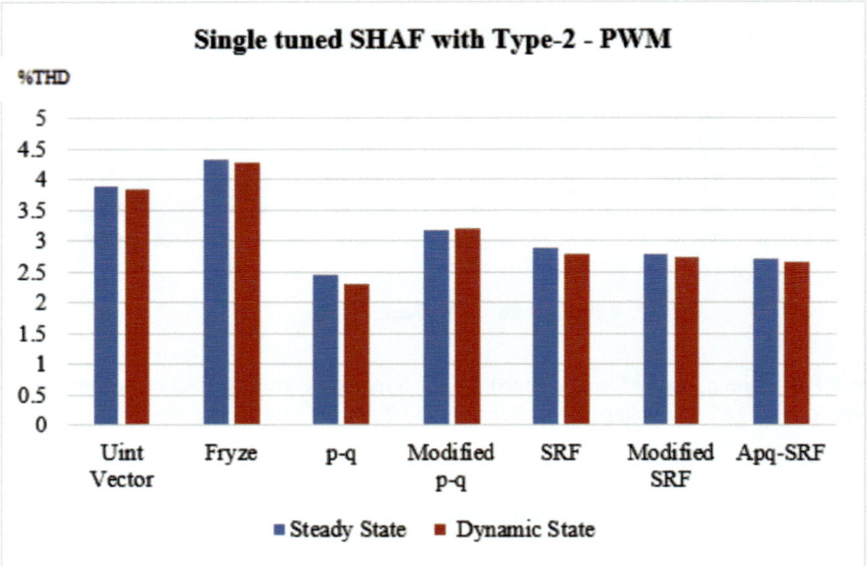

Figure 4.133 Comparison of single-tuned SHAF reference current extraction techniques with type-2 fuzzy DC-link voltage controller and PWM pulse generator.

has been observed that, all the reference current extraction techniques are able to cancel out grid current harmonics and excess reactive power as per the IEE-519-2022 standards. The single-tuned SHAF with proposed Apq-SRF reference current extraction technique is able to cancel out grid current harmonics and excess reactive power in conjunction with type-2 fuzzy DC capacitor voltage controller and PWM pulse generator is as shown in Figure 4.133.

The performance comparison of various reference current extraction techniques with type-2 fuzzy DC capacitor voltage regulator and PWM pulse generator for double-tuned SHAF is as shown in Figure 4.134. It has been observed that, all the reference current extraction techniques are able to cancel out grid current harmonics and excess reactive power as per the IEE-519-2022 standards. The double-tuned SHAF with proposed Apq-SRF reference current extraction technique is able to cancel out grid current harmonics and excess reactive power in conjunction with type-2 fuzzy DC capacitor voltage controller and PWM pulse generator is as shown in Figure 4.134.

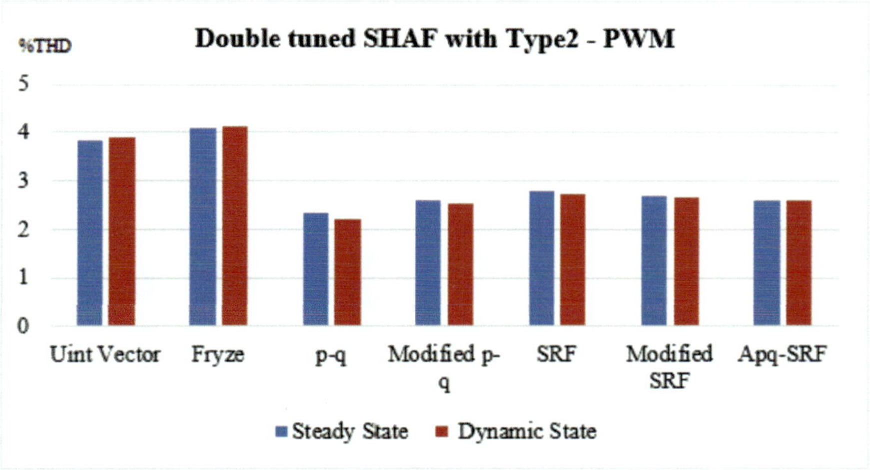

Figure 4.134 Comparison of double-tuned SHAF reference current extraction techniques with type-2 fuzzy DC-link voltage controller and PWM pulse generator.

4.9.12 Comparison of various reference current extraction techniques with type-2 fuzzy DC-link voltage controller and HCC pulse generator

The performance comparison of various reference current extraction techniques with type-2 fuzzy DC capacitor voltage regulator and HCC pulse generator for SAF is as shown in Figure 4.135. It has been observed that, all the reference current extraction techniques are able to cancel out grid current harmonics and excess reactive power as per the IEE-519-2022 standards. The SAF with proposed Apq-SRF reference current extraction technique is able to cancel out grid current harmonics and excess reactive power in conjunction with type-2 fuzzy DC capacitor voltage controller and HCC pulse generator is as shown in Figure 4.135.

The performance comparison of various reference current extraction techniques with type-2 fuzzy DC capacitor voltage regulator and HCC pulse generator for single-tuned SHAF is as shown in Figure 4.136. It has been observed that, all the reference current extraction techniques are able to cancel out grid current harmonics and excess reactive power as per the IEE-519-2022 standards. The single-tuned SHAF with proposed Apq-SRF reference current extraction technique is able to cancel out grid current harmonics and excess reactive power in conjunction with type-2 fuzzy DC capacitor voltage controller and HCC pulse generator is as shown in Figure 4.136.

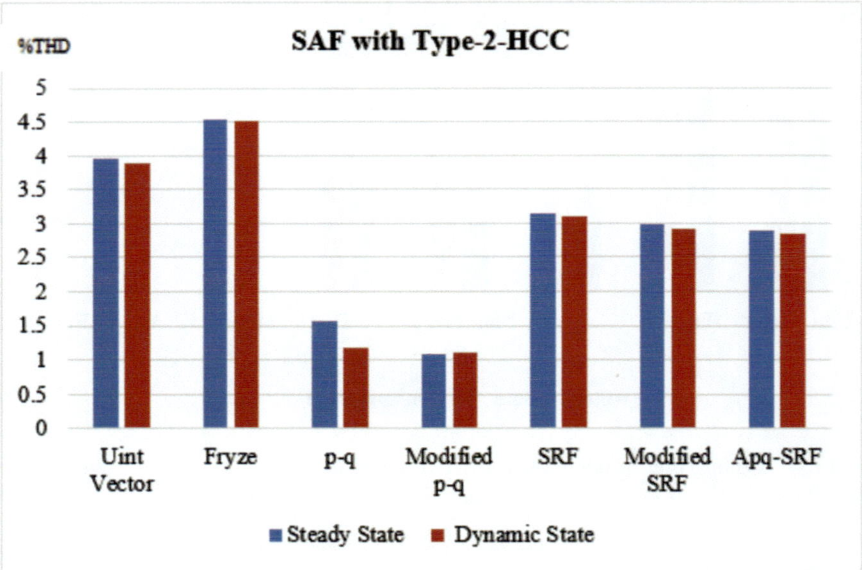

Figure 4.135 Comparison of SAF reference current extraction techniques with type-2 fuzzy DC-link voltage controller and HCC pulse generator.

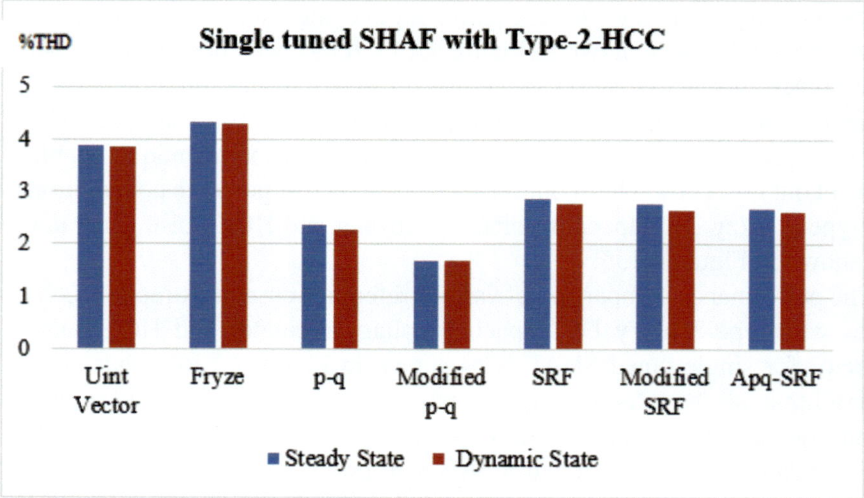

Figure 4.136 Comparison of single-tuned SHAF reference current extraction techniques with type-2 fuzzy DC-link voltage controller and HCC pulse generator.

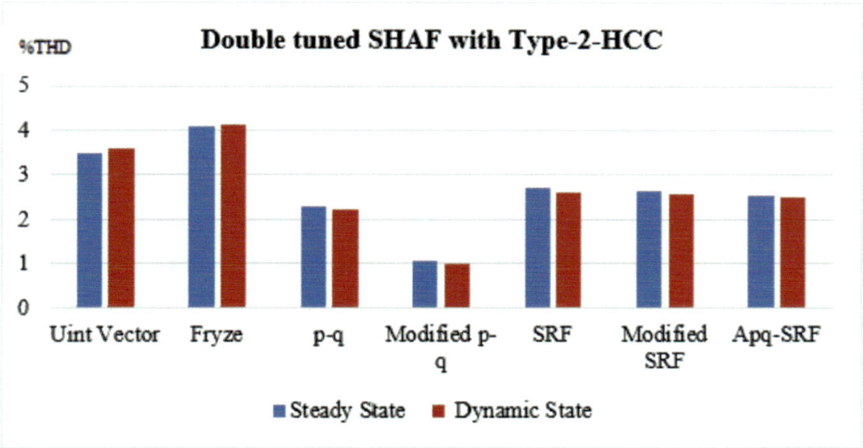

Figure 4.137 Comparison of double-tuned SAF reference current extraction techniques with type-2 fuzzy DC-link voltage controller and HCC pulse generator.

The performance comparison of various reference current extraction techniques with type-2 fuzzy DC capacitor voltage regulator and HCC pulse generator for double-tuned SHAF is as shown in Figure 4.137. It has been observed that, all the reference current extraction techniques are able to cancel out grid current harmonics and excess reactive power as per the IEE-519-2022 standards. The double-tuned SHAF with proposed Apq-SRF reference current extraction technique is able to cancel out grid current harmonics and excess reactive power in conjunction with type-2 fuzzy DC capacitor voltage controller and HCC pulse generator is as shown in Figure 4.137.

4.9.13 Comparison of various reference current extraction techniques with type-2 fuzzy DC-link voltage controller and AHCC pulse generator

The performance comparison of various reference current extraction techniques with type-2 fuzzy DC capacitor voltage regulator and AHCC pulse generator for SAF is as shown in Figure 4.138. It has been observed that, all the reference current extraction techniques are able to cancel out grid current harmonics and excess reactive power as per the IEE-519-2022 standards. The SAF with proposed Apq-SRF reference current extraction technique is able to cancel out grid current harmonics and excess reactive power in conjunction with type-2 fuzzy DC capacitor voltage controller and AHCC pulse generator is as shown in Figure 4.138.

The performance comparison of various reference current extraction techniques with type-2 fuzzy DC capacitor voltage regulator and AHCC pulse generator for single-tuned SHAF is as shown in Figure 4.139. It has been observed that, all the reference current extraction techniques are able to cancel out grid current harmonics and excess reactive power as per the IEE-519-2022 standards. The single-tuned SHAF with proposed Apq-SRF reference current extraction technique is able to cancel out grid current harmonics and excess reactive power in conjunction with type-2 fuzzy DC capacitor voltage controller and AHCC pulse generator is as shown in Figure 4.139.

The performance comparison of various reference current extraction techniques with type-2 fuzzy DC capacitor voltage regulator and AHCC pulse generator for double-tuned SHAF is as shown in Figure 4.140. It has been observed that, all the reference current extraction techniques are able to cancel out grid current harmonics and excess reactive power as per the IEE-519-2022 standards. The double-tuned SHAF with proposed

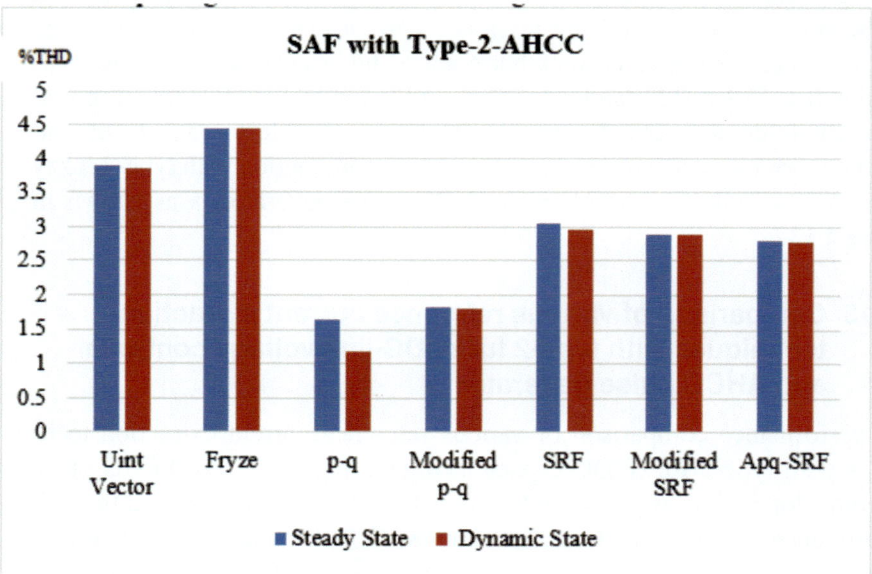

Figure 4.138 Comparison of SAF reference current extraction techniques with type-2 fuzzy DC-link voltage controller and AHCC pulse generator.

Apq-SRF reference current extraction technique is able to cancel out grid current harmonics and excess reactive power in conjunction with type-2 fuzzy DC capacitor voltage controller and AHCC pulse generator is as shown in Figure 4.140.

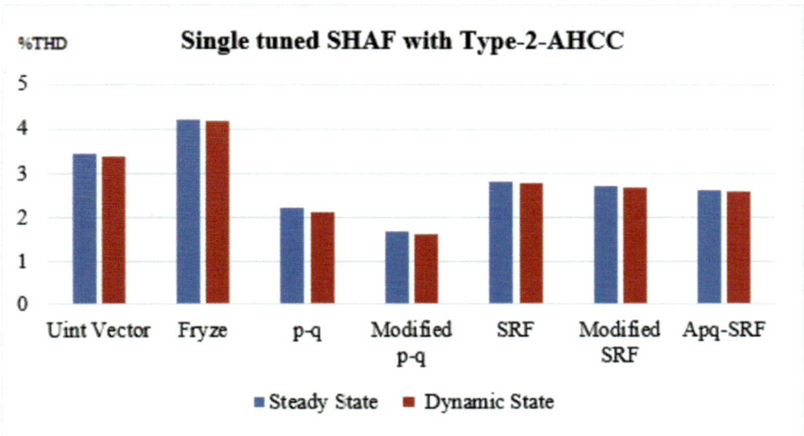

Figure 4.139 Comparison of single-tuned SHAF reference current extraction techniques with type-2 fuzzy DC-link voltage controller and AHCC pulse generator.

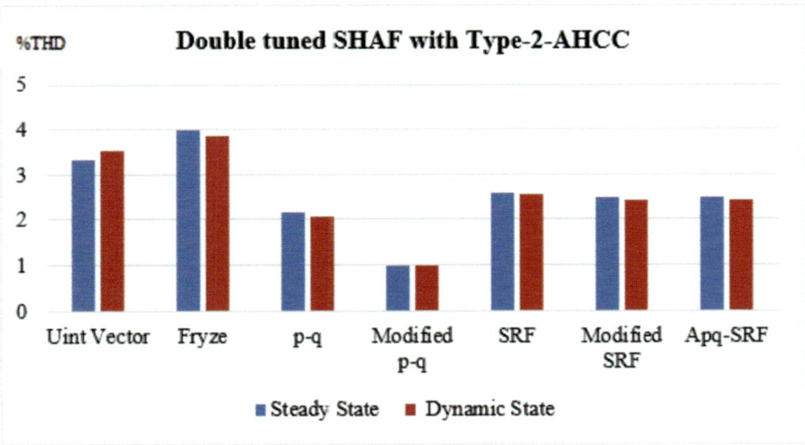

Figure 4.140 Comparison of double-tuned SHAF reference current extraction techniques with type-2 fuzzy DC-link voltage controller and AHCC pulse generator.

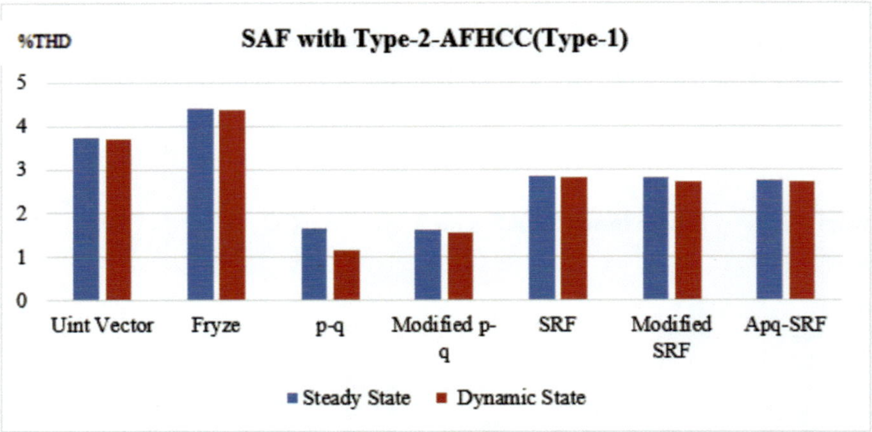

Figure 4.141 Comparison of SAF reference current extraction techniques with type-2 fuzzy DC-link voltage controller and type-1 AFHCC pulse generator.

4.9.14 Comparison of various reference current extraction techniques with type-2 fuzzy DC-link voltage controller and type-1 AFHCC pulse generator

The performance comparison of various reference current extraction techniques with type-2 fuzzy DC capacitor voltage regulator and type-1 AFHCC pulse generator for SAF is as shown in Figure 4.141. It has been observed that, all the reference current extraction techniques are able to cancel out grid current harmonics and excess reactive power as per the IEE-519-2022 standards. The SAF with proposed Apq-SRF reference current extraction technique is able to cancel out grid current harmonics and excess reactive power in conjunction with type-2 fuzzy DC capacitor voltage controller and type-1AFHCC pulse generator is as shown in Figure 4.141.

The performance comparison of various reference current extraction techniques with type-2 fuzzy DC capacitor voltage regulator and type-1 AFHCC pulse generator for single-tuned SHAF is as shown in Figure 4.142. It has been observed that, all the reference current extraction techniques are able to cancel out grid current harmonics and excess reactive power as per the IEE-519-2022 standards. The single-tuned SHAF with proposed Apq-SRF reference current extraction technique is able to cancel out grid current harmonics and excess reactive power in conjunction with type-2 fuzzy DC capacitor voltage controller and type-1 AFHCC pulse generator is as shown in Figure 4.142.

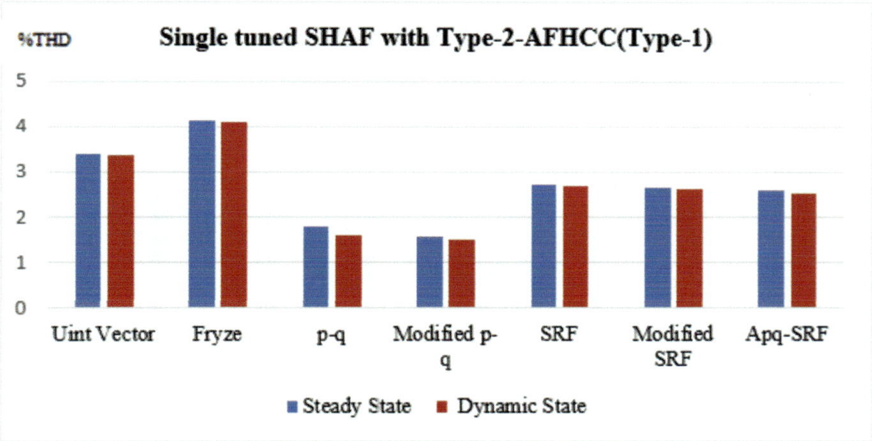

Figure 4.142 Comparison of single-tuned SHAF reference current extraction techniques with type-2 fuzzy DC-link voltage controller and type-1 AFHCC pulse generator.

The performance comparison of various reference current extraction techniques with type-2 fuzzy DC capacitor voltage regulator and type-1 AFHCC pulse generator for double-tuned SHAF is as shown in Figure 4.143. It has been observed that, all the reference current extraction techniques are able to cancel out grid current harmonics and excess reactive power as per the IEE-519-2022 standards. The double-tuned SHAF with proposed Apq-SRF reference current extraction technique is able to cancel out grid current harmonics and excess reactive power in conjunction with type-2 fuzzy DC capacitor voltage controller and type-1 AFHCC pulse generator is as shown in Figure 4.143.

4.9.15 Comparison of various reference current extraction techniques with type-2 fuzzy DC-link voltage controller and type-2 AFHCC pulse generator

The performance comparison of various reference current extraction techniques with type-2 fuzzy DC capacitor voltage regulator and type-2 AFHCC pulse generator for SAF is as shown in Figure 4.144. It has been observed that, all the reference current extraction techniques are able to cancel out grid current harmonics and excess reactive power as per the IEE-519-2022 standards. The SAF with proposed Apq-SRF reference current extraction technique is able to cancel out grid current harmonics and excess reactive

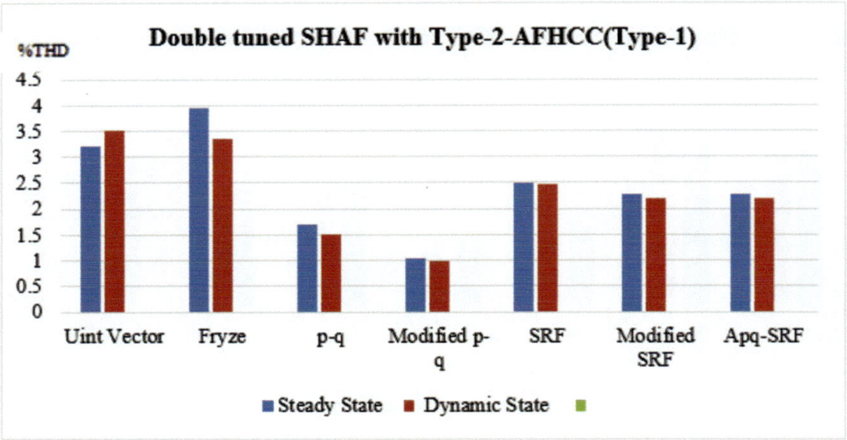

Figure 4.143 Comparison of double-tuned SHAF reference current extraction techniques with type-2 fuzzy DC-link voltage controller and type-1 AFHCC pulse generator.

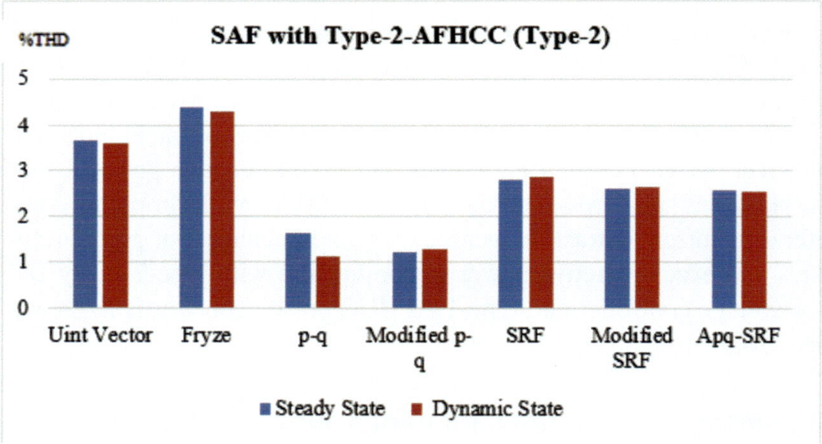

Figure 4.144 Comparison of SAF reference current extraction techniques with type-2 fuzzy DC-link voltage controller and type-2 AFHCC pulse generator.

power in conjunction with type-2 fuzzy DC capacitor voltage controller and type-2 AFHCC pulse generator is as shown in Figure 4.144.

The performance comparison of various reference current extraction techniques with type-2 fuzzy DC capacitor voltage regulator and type-2 AFHCC pulse generator for single-tuned SHAF is as shown in Figure 4.14. It has

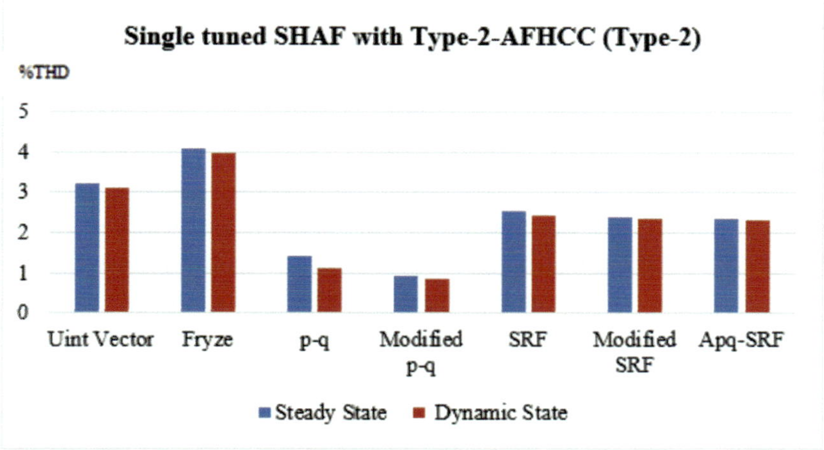

Figure 4.145 Comparison of single-tuned SHAF reference current extraction techniques with type-2 fuzzy DC-link voltage controller and type-2 AFHCC pulse generator.

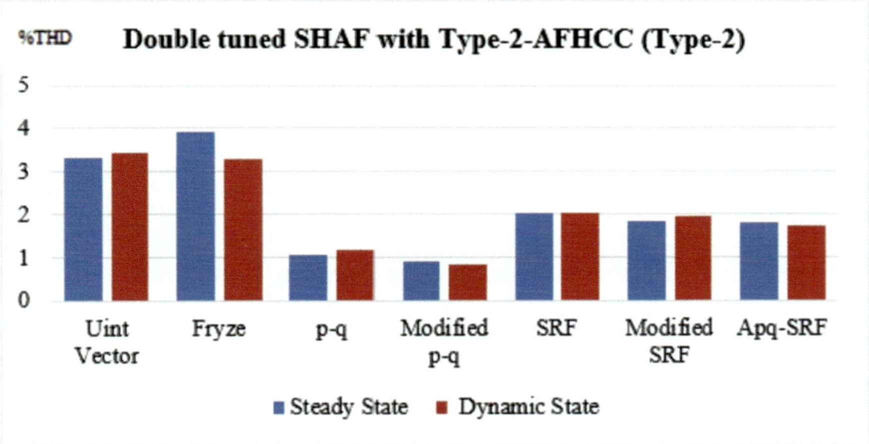

Figure 4.146 Comparison of double-tuned SHAF reference current extraction techniques with type-2 fuzzy DC-link voltage controller and type-2 AFHCC pulse generator.

been observed that, all the reference current extraction techniques are able to cancel out grid current harmonics and excess reactive power as per the IEE-519-2022 standards. The single-tuned SHAF with proposed Apq-SRF reference current extraction technique is able to cancel out grid current harmonics and excess reactive power in conjunction with type-2 fuzzy DC capacitor voltage controller and type-2 AFHCC pulse generator is as shown in Figure 4.145.

The performance comparison of various reference current extraction techniques with type-2 fuzzy DC capacitor voltage regulator and type-2 AFHCC pulse generator for double-tuned SHAF is as shown in Figure 4.146. It has been observed that, all the reference current extraction techniques are able to cancel out grid current harmonics and excess reactive power as per the IEE-519-2022 standards. The double-tuned SHAF with proposed Apq-SRF reference current extraction technique is able to cancel out grid current harmonics and excess reactive power in conjunction with type-2 fuzzy DC capacitor voltage controller and type-2 AFHCC pulse generator is as shown in Figure 4.146.

4.10 Contributions

1. The overall comparison and analysis of the SAF existing and proposed reference current extraction techniques, DC-link voltage controllers and pulse-generating techniques have been performed using MATLAB.
2. The comparison of the SAF control techniques with respect to the two single-tuned passive filters or one double-tuned passive filter has been performed in MATLAB.
3. All the MATLAB results are compared with IEE-519-2022 standards.

4.11 Summary

In this chapter, the SHAF is performed using MATLAB Simulink. Various DC capacitor voltage controllers like PI/type-1 fuzzy/type-2 fuzzy, different reference current extraction techniques such as unit vector/Fryze power/p-q theory/modified p-q theory/SRF/modified SRF/proposed Apq-SRF, and various pulse generators of PWM/HCC/AHCC/AFHCC type-1/AFHCC type-2 have been applied to evaluate the performance of SAF. Similarly, STPF and DTPF also have been modeled and their performance is evaluated in conjunction with various SAF control techniques. Therefore, the whole SHAF performance is evaluated with various SAF control techniques and tuned passive filters. These simulation results make known that the SHAF is cancelling out current harmonics as well as excess reactive power. The comparison results of the THD, real and reactive power measurements are presented in the tabular form. The measured THD of the source currents is less than 5% in all cases, which follows IEE-519-2022 standards.

5

Conclusions and Future Scope

5.1 Conclusions

Combining the SAF with two single-tuned or one double-tuned passive filters per phase, the SHAF is intended to eliminate power factor, extra reactive power, and current harmonics in the distribution network.

A variety of reference current extraction methods, PWM switching systems, tunable passive filters, and controllers for preserving a steady VSI capacitor voltage are all part of the SHAF. Essentially a VSI, the SAF generates pulses using a variety of switching methods, including PWM, HCC, AHCC, AFHCC type-1, and AFHCC type-2. The following conclusions are reached after MATLAB simulations and TMS5700-XL hardware are used to execute and investigate the various reference current extraction techniques, pulse generation techniques, techniques for constant VSI capacitor voltage, and tuned passive filters of single-tuned/double-tuned for SHAF.

The DC voltage of the VSI is controlled using the PI/type-1 fuzzy/type-2 fuzzy approaches to predict the maximum value of the reference current. These techniques for managing DC voltages with various pulse-generating technologies are being studied for loads of uncontrolled diode bridge rectifiers.

However, the suggested network is difficult to obtain under parametric fluctuations, non-linearity, and load disturbances, and the PI controller demands a precise mathematical computation of it all. The optimization of SHAF performance is therefore quite challenging. Fuzzy logic, either type 1 or type 2, is used to overcome the drawbacks of PI controllers since it does not necessitate precise mathematical calculations and can operate with imperfect inputs.

The modified p-q theory incorporates the traditional p-q theory notion with a positive sequence voltage detector. Uncontrolled diode bridge rectifier loads are used to verify the sinusoidal extraction controller-based SAF system using several direct pulse generating approaches. Nevertheless, it takes longer to calculate than traditional p-q control theory.

Fryze power theory has the advantage of working directly with line currents and *abc*-phase voltage, which lessens the workload associated with calculating reference currents. This control method is very easy to implement because it does not require the usage of Clarke transformation. This theory's drawback is that it lacks a consistent equation for calculating the three-phase instantaneous reactive power, which is needed to offset the reactive power applied to the load.

The modified SRF theory uses simple and efficient unit vector instead of PLL circuit. The modified SRF theory ensures the better performance than conventional SRF method in terms of V_{dc}-settling time, harmonics, and reactive-power balance.

A new Apq-SRF reference current extraction method is created that eliminates the drawbacks of traditional p-q and SRF control. The advantages of traditional p-q and SRF theory will be extracted by Apq-SRF control, which also eliminates their drawbacks.

When compared to two single-tuned passive filters, the double-tuned passive filter has demonstrated superior performance. Additionally, it has the benefit of just requiring one circuit breaker per phase, as opposed to two circuit breakers per phase for two single-tuned passive filters. Compared to the analogous method, the parametric method-based double-tuned passive filter has shown superior performance.

These results show that the SHAF always removes power factor, excess reactive power, and current harmonics. The proposed Apq-SRF method has been shown to be robust to all supply and load conditions. The type-2 AFHCC switching technique enhances SHAF performance and lowers switching power losses when compared to PWM, HCC, AHCC, and AFHCC type-1. Grid current THD consistently meets the harmonic standards of IEEE 519-2022.

5.2 Future Scope

A complete Field programmable gate array (FPGA) or Real time digital simulator (RTDS) or D-Space-based controller for the development of SHAF would increase controller design freedom. The amount of hardware needed would be greatly decreased.

Recently, renewable energy sources have gained significant attention, and in many instances, these power sources are supplying a grid that can also support non-linear loads. For such systems, an integrated SAF would be necessary. For additional research, a full model incorporating the grid, SAF, and renewable energy sources may be adopted. Multi-level inverters would be required to accommodate higher power requirements; it is recommended that the SAF adopt such inverters and control techniques.

From an inverter's frequency of operation standpoint, reactive-power compensation and harmonic compensation are basically two distinct requirements. A high-frequency low-power inverter is necessary for harmonic compensation, while low-frequency high-power operation is necessary for reactive power. In order to decouple these duties and enable two separate inverters to perform compensation, it is crucial to design an appropriate approach.

SAFs based on model predictive controllers (MPCs) have demonstrated a high level of effectiveness in reducing current harmonics. A cost function and a predictor are both part of the MPC. With minimal switching losses, the cost function maximizes SAF performance while precisely tracking both the reference current and the SAF compensatory current. PWM switching techniques, an external emulator, are not necessary for the SAF functioning when using the finite control set-MPC (FCS-MPC) methodology.

The distribution network has recently been equipped with triple-tuned passive filters to reduce three lower-order current harmonics. It performs better than two passive filters that are single-tuned and double-tuned. The difficult task is that it requires a lot of mathematical calculations and has a sophisticated design.

Multi-level inverters with reduced switches can be used as a VSI in the SAF. The performance of these is superior as compared with conventional inverters.

Bibliography

[1] Bollen, M.H.J."Understanding Power Quality Problems, Voltage Sags and Interruptions". *New York: IEEE Press*, 2000.

[2] Grady, W. Mack, and Surya Santoso. "Understanding power system harmonics." *IEEE Power Engineering Review*, pp.8-11, 2001.

[3] G. T. Heydt, "Electric Power Quality," *West Lafayette, IN: Stars in a Circle*, 1991.

[4] A. Moreno-Munoz, "Power Quality: Mitigation Technologies in a Distributed Environment" *London, U.K.: Springer-Verlag*, 2007.

[5] N. Mohan, T. M. Undeland and W. P. Robbins, "Power Electronics: Converters Applications and Design", *New York, USA:John Wiley & Sons Inc.,* 1995.

[6] Recommended Practices and Requirements for Harmonic Control in Electric Power Systems, *IEEE Standard 519-1992, New York*, 1993.

[7] IEEE Recommended Practice for Evaluating Electric Power System Compatibility With Electronic Process Equipment, *IEEE std.* 1346-1998.

[8] "IEEE Recommended Practice and Requirements for Harmonic Control in Electric Power Systems," in *IEEE Std 519-2014 (Revision of IEEE Std 519-1992)* , vol., no., pp.1-29, 11 June 2014, doi: 10.1109/IEEESTD.2014.6826459

[9] A. Baitha and N. Gupta, "A comparative analysis of passive filters for power quality improvement." *2015 International Conference on Technological Advancements in Power and Energy (TAP Energy), Kollam, India*, 2015, pp. 327-332, doi: 10.1109/TAPENERGY.2015.7229640

[10] J.C. Das, "Passive filters-potentialities and limitations", *IEEE Trans. on Industry Applications*, vol. 40, no. 1, pp. 232-241, Jan./Feb. 2004.

[11] K. R. Cheepati, E. Parimalasundar, K. Suresh, C. R. Reddy, M. M. Alqahtani and M. Khalid, "Design of Triple Tuned Passive Harmonic Power Filter—A Novel Approach," in IEEE Canadian Journal of Electrical and Computer Engineering, vol. 46, no. 4, pp. 270-277, Fall 2023, doi: 10.1109/ICJECE.2023.3296826.

[12] Cheepati, Kumar Reddy, Sardar Ali, and M. Surya Kalavathi. "Performance analysis of double tuned passive filter for power quality." International Journal of Control Theory and applications 9.7 (2016): 3295-3305.

[13] S.N. A. L. Yousif, M. Z. C. Wanik and A. Mohamed, "Implementation of Different Passive Filter Designs for Harmonic Mitigation", *Proc. of National Power & Energy Conference (PECon)*, pp. 229-234, 2004.

[14] N. Mohan and G. R. Kamath, "A novel, high-frequency, per-phase static VAr compensator," *IAS '95. Conference Record of the 1995 IEEE Industry Applications Conference Thirtieth IAS Annual Meeting, Orlando, FL, USA*, 1995, pp. 2581-2586 vol.3, doi: 10.1109/IAS.1995.530631.

[15] A.H.Noyola W.M.Grady, M.J.Samotyj, "Survey of Active Power Line Conditioning Methodologies", *IEEE Tran. on Power Deliv.,* vol. 5, pp. 1536–1542, 1990.

[16] An Luo, Zhikang Shuai, Wenji Zhu, Ruixiang Fan and Chunming Tu "Development of Hybrid Active Power Filter Based on the Adaptive Fuzzy Dividing Frequency-Control Method*" IEEE transactions on power delivery*, vol. 24, no. 1, pp.424-432, Jan., 2009.

[17] Salem Rahmani, Member, AbdelhamidHamadi,, and Kamal Al-Haddad, "A Lyapunov-Function-Based Control for a Three-Phase Shunt Hybrid Active Filter" *IEEE transactions on industrial electronics*, vol. 59, no. 3, March, 2012.

[18] Victor Fabián Corasaniti, Maria Beatriz Barbieri, , Patricia LilianaArnera and María Inés Valla "Hybrid Power Filter to Enhance Power Quality in a Medium-Voltage Distribution Network" *IEEE transactions on industrial electronics*, vol. 56, no. 8, August, 2009.

[19] Salem Rahmani, AbdelhamidHamadi,, NassarMendalek, and Kamal Al-Haddad "A New Control Technique for Three-Phase Shunt Hybrid Power Filter" *IEEE transactions on industrial electronics*, vol. 56, no. 8, August, 2009.

[20] Cheepati, Kumar Reddy, Nageswara Rao Maddala, and Surya Kalavathi Munagala. "A Novel Reference Current Extraction Technique with Multi-Functional Capability for Shunt Active Filter." Journal of Electrical Engineering & Technology 15.2 (2020): 657-672.

[21] Cheepati, Kumar Reddy, A. L. I. Sardar, and Surya Kalavathi Munagala. "Double tuned series hybrid active power filter for power quality improvement in distribution system under non linear load conditions." UPB Sci. Bull., Series C 79.1 (2017): 245-260.

[22] A. Baitha and N. Gupta, "A comparative analysis of passive filters for power quality improvement," *2015 International Conference on Technological Advancements in Power and Energy (TAP Energy)*, Kollam, India, 2015, pp. 327-332, doi: 10.1109/TAPENERGY.2015.7229640.

[23] Yap Hoon , Mohd Amran Mohd Radzi, Muhammad Ammirrul Atiqi Mohd Zainuri and Mohamad Adzhar Md Zawawi "Shunt Active Power Filter: A Review on Phase Synchronization Control Techniques" *Electronics*, vol.8, no.791, pp.1-20, 2019.

[24] Cheepati, Kumar Reddy, Nageswara Rao Maddala, and Surya Kalavathi Munagala. "A hybrid reference current extraction technique with multifunctional capability for shunt active filter in all-electric aircraft." International Journal of Ambient Energy 43.1 (2022): 2432-2445.

[25] M.El-Habrouk and M.K.Darwish "Design and implementation of a modified Fourier analysis harmonic current computation technique for power active filters using DSPs" *IEE Proceedings*, pp.21-28, April, 2000.

[26] Cheepati, Kumar Reddy. "Analysis of Shunt Active Filter Performance under Different Supply and Loading Conditions." Advanced Control & Optimization Paradigms for Energy System Operation and Management. River Publishers, 2023. 165-189.

[27] Cheepati, Kumar Reddy, et al. "A two-step horizon optimum switching vector-model predictive control with a novel shunt active filter reference current extraction." Journal of Circuits, Systems and Computers 31.10 (2022): 2250174.

[28] Manish Kr. Singh and Jayanti Choudhary "Fryze Current Controller Based Active Power Filter", *IFAC Proceedings*, vol.17, no.1, pp. 794-800, 2014.

[29] Edson H.Watanabe, Hirofumi Akagi and Maurício Aredes "The P-Q theory for Active Filter Control: Some Problems and Solutions" *Revista Controle & Automação*,vol.15, no.1, pp.78-84, 2004 .

[30] Bhattacharjee, Kakoli "Design and Simulation of Synchronous Reference Frame Based Shunt Active Power Filter Using SIMULINK". *IET Conference Publications*, 2013.

[31] João Afonso, Maurício Aredes, Edson Watanabe and Júlio Martins "Shunt Active Filter for Power Quality Improvement" *International Conference UIE 2000 – "Electricity for a Sustainable Urban Development" Lisboa, Portugal*, pp. 683-691, 2000.

[32] Peng FZ, Lai J1S. "Generalized Instantaneous Reactive Power Theory for Three Phase Power Systems" *IEEE Trans. on Instrument and Measurement*, 1996.

[33] Herrera, Reyes & Revuelta, Patricio," Instantaneous Reactive Power Theory: A Comparative Evaluation of Different Formulations" *Power Delivery, IEEE Transactions*, vol. 22, pp.595 – 604, 2007.

[34] Rondineli Rodrigues Pereira, Carlos Henrique da Silva "New Strategies for Application of Adaptive Filters in Active Power Filters" *IEEE Transactions on Industry Applications*, vol. 47, no. 3, may/june, 2011.

[35] R. L. de Araujo Ribeiro, C. C. de Azevedo and R. M. de Sousa, "A Robust Adaptive Control Strategy of Active Power Filters for Power-Factor Correction, Harmonic Compensation, and Balancing of Nonlinear Loads," in *IEEE Transactions on Power Electronics*, vol. 27, no. 2, pp. 718-730, Feb. 2012, doi: 10.1109/TPEL.2011.2161334.

[36] Kazmierkowski, M.P. & Malesani, Luigi "Current control techniques for three-phase voltage-source PWM converters: A survey". *Industrial Electronics, IEEE Transactions* vol. 45. pp.691 - 703.

[37] Akira Nabae, Satoshi Ogasawara, and Hirofumi Akagi, "A Novel Control Scheme for Current-Controlled PWM Inverters" *IEEE Trans. on Industry Applications*, vol no. 22, no.4, pp.562-570, July,1986.

[38] Bimal K. Bose "An Adaptive Hysteresis-Band Current Control Technique of a Voltage-Fed PWM Inverter for Machine Drive System, " *IEEE Transaction on Industrial Electronics*, vol.37, no.5, October, 1990.

[39] J. W. Dixon, S. Tepper, and L. Moran, "Analysis and evaluation of different modulation techniques for active power filters," *in Proc. IEEEAPEC'94*, pp. 894–900, 1994.

[40] Bong-Hwan Kwon, Tae-Woo Kim and Jang-Hyoun Youm, "A novel SVM-based hysteresis current controller," in *IEEE Transactions on Power Electronics*, vol. 13, no. 2, pp. 297-307, March, 1998, doi: 10.1109/63.662844.

[41] G. H. Bode and D. G. Holmes, "Implementation of three level hysteresis current control for a single phase voltage source inverter," *2000 IEEE 31st Annual Power Electronics Specialists Conference. Conference Proceedings (Cat. No.00CH37018)*, Galway, Ireland, 2000, pp. 33-38 vol.1, doi: 10.1109/PESC.2000.878796.

[42] Kale, Murat & Ozdemir, Engin "An adaptive hysteresis band current controller for shunt active power filter" *Electric Power Systems Research - ELEC POWER SYST RES*, vol.73, pp.113-119, 2005.

[43] Hamoudi, F., Chaghi, A., Amimeur, H., and Merabet, E., "Adaptive Fuzzy Hysteresis Band Current Controller for Four-Wire Shunt Active Filter", *in Intelligent Systems and Automation*, 2008, vol. 1019, pp. 197–202. doi:10.1063/1.2952978.

[44] B. Mazari and F. Mekri, "Fuzzy Hysteresis Control and Parameter Optimization of a Shunt Active Power Filter", *Journal of Information Science and Engineering*, vol.21, pp.1139-1156, 2005.

[45] Suresh Mikkili, A. K. Panda, "Real-time implementation of PI and Fuzzy logic controllers based shunt active filter control strategies for power quality improvement," International Journal of Electrical Power and Energy Systems – Elsevier, vol.43, pp.1114-1126, 2012.

[46] Bhim Singh, A.Chandra, K. Al-haddad "Power quality problems and mitigation techniques" *wiley, 1st edition*, 2015.

[47] S. Rechka, E. Ngandui, J. Xu and P. Sicard, "A comparative study of harmonic detection algorithms for active filter and hybrid active filers", *IEEE Power Elec. Specialist Conf.*, pp. 357-363, 2002.

[48] Roger C. Dugan, Mark, F. McGranaghan, Surya Santoso and H.Wayne Beaty "Electrical Power Systems Quality" *Tata Mcgrw-hill 3^{rd}edition*, 2013.

[49] Young-Sik Cho and Hanju Cha "Single-tuned Passive Harmonic Filter Design Considering Variances of Tuning and Quality Factor", *Journal of International Council on Electrical Engineering*, vol.1, no.1, pp.7-13, 2011.

[50] H. Yi-hong and SU Heng "A New Method of Designing Double Tuned Filter" *Proceedings of the 2nd International Conference on Computer Science and Electronics Engineering*, 2013).

[51] M.A. Zamani and M.Moshneni, "Damped type Double Tuned Filters Design for HVDC Systems", *9th International Conference, Electrical Power Quality and Utilization, Barcelona*, 2007.

[52] X. Yao, "Algorithm for the Parameters of Double Tuned Filter", *IEEE International Conference on Harmonic and Quality of Power, Greece*, 1998.

[53] K.R. Cheepati, Ali Sardar, M. Kalavathi," Performance Analysis of Double Tuned Passive Filter for Power Quality", *International Journal of Control Theory and applications*, vol. 9, no.7, pp. 3295-3305, 2016.

[54] K.R. Cheepati, Ali Sardar, M. Kalavathi,"Overview of Double Tuned Harmonic Filters in Improving Power Quality under Non Linear Load Conditions", *International Journal of Grid and Distributed Computing*, vol.10, no.7, pp.11-26, 2017.

[55] L. A Moran, J. W. Dixon, "Active Filters", Chapter 33 in "Power Electronics Handbook", *Academic Press*, pp. 829-841, August 2001.

[56] G. Deep Srivastava and R. D. Kulkarni, "Design, simulation and analysis of Shunt Active Power Filter using instantaneous reactive power topology," *2017 International Conference on Nascent Technologies in Engineering (ICNTE)*, Navi Mumbai, 2017, pp. 1-6, doi: 10.1109/ICNTE.2017.7947937.

[57] Bhim Singh, Kamal Al-Haddad and Ambrish Chandra, "A Review of Active Filters for Power Quality Improvement" *IEEE Trans. Ind. Applicat.*, vol. 46, no.5 pp. 960-971, October, 1999.

[58] https://www.electronicshub.org/pid-controller-working-and-tuning-methods/

[59] Suresh Mikkili and Anup Kumar Panda "Types-1 and -2 fuzzy logic controllers-based shunt active filter Id–Iq control strategy with different fuzzy membership functions for power quality improvement using RTDS hardware" *IET Power Electron.,* vol. 6, no. 4, pp. 818–833, 2013.

[60] M. Aredes and L. F. C. Monteiro, "A control strategy for shunt active filter," *10th International Conference on Harmonics and Quality of Power. Proceedings (Cat. No.02EX630), Rio de Janeiro, Brazil*, 2002, pp. 472-477 vol.2, doi: 10.1109/ICHQP.2002.1221481.

[61] G. S. Raj and K. Rathi, "P-Q theory based Shunt Active Power Filter for power quality under ideal and non-ideal grid voltage conditions," *2015 International Conference on Power, Instrumentation, Control and Computing (PICC), Thrissur, India*, 2015, pp. 1-5, doi: 10.1109/PICC.2015.7455754.

[62] S. Sahadev and Manju B, "Performance analysis of SRF based SAPF with PI and FUZZY controllers," *2015 International Conference on Control Communication & Computing India (ICCC), Trivandrum, India*, 2015, pp. 260-264, doi: 10.1109/ICCC.2015.7432902.

[63] Suleiman Musa, Mohd Amran Mohd Radzi, Hashim Hizam , Noor Izzri Abdul Wahab, Yap Hoon and Muhammad Ammirrul Atiqi Mohd Zainuri "Modified Synchronous Reference Frame Based Shunt Active Power Filter with Fuzzy Logic Control Pulse Width Modulation Inverter" *Energies* vol.10, no.758, pp.1-16, 2017.

[64] H. Golwala and R. Chudamani, "Comparative study of switching signal generation techniques for three phase four wire shunt active power filter," *2011 IEEE International Electric Machines & Drives Conference (IEMDC), Niagara Falls, ON, Canada*, 2011, pp. 1409-1414, doi: 10.1109/IEMDC.2011.5994814.

[65] Biricik, Samet, and Hasan Komurcugil. "Three-level hysteresis current control strategy for three-phase four-switch shunt active filters." *IET Power Electronics* vol.9, no.8, pp.1732-1740 2016.

[66] Kale, Murat, and Engin Ozdemir. "A novel adaptive hysteresis band current controller for shunt active power filter." *Proceedings of 2003 IEEE Conference on Control Applications*, 2003, vol. 2, 2003.

[67] Suresh, Y., A. K. Panda, and M. Suresh. "Real-time Implementation of Adaptive Fuzzy Hysteresis-Band Current Control Technique for Shunt Active Power Filter." *IET Power Electronics,*vol.5, no.7, pp.1188-1195, 2012.

Index

About the Authors

Kumar Reddy Cheepati obtained his Ph.D. degree from JNTUK, Kakinada in 2021. Currently he is working as an Associate Professor in the department of Electrical and Electronics Engineering at KSRM College of Engineering (Autonomous), Kadapa, Andhra Pradesh. He has published research papers in various international journals of high repute which include Scopus, SCI and ESCI indexed journals. He attended and presented papers in prestigious international conferences which include IEEE. Dr. Reddy is mainly working on predictive control techniques for shunt active filters to mitigate the harmonics in distribution networks as per IEEE-519-2022 standards. He is an active reviewer of the Electrical Power System Research (EPSR) journal, Journal of Circuits, Systems and Computers (JCSC), Journal of Engineering Research (JER) and Circuit World journal. He has been invited to several research meetings, conferences and guest lectures.

T. Mariprasath received a Ph.D. degree from the Rural Energy Centre at The Gandhigram Rural Institute (deemed to be a university) in January 2017. Since June 2018, he has been working as an Associate Professor in the Department of EEE at KSRM College of Engineering (Autonomous), Andhra Pradesh, India. He is also a member of the R&D Cell. He has published eleven journal articles in the Science Citation Index and fifteen articles in Scopus. He received a grant from the Ministry of Micro, Small, and Medium Enterprises to develop a self-powered GPS tracker. Springer Discover Electronics recently selected him as an editorial board member. He serves as a reviewer for reputed publishers like IEEE, Elsevier, Wiley, Springer, Taylor & Francis, and IET. His research interests include electric vehicles, solar PV, IoT, machine learning, and green materials.

Marco Esteban Rivera Abarca received an Electronic Civil Engineering degree and M.Sc. degree in Engineering, with specialization in Electrical Engineering, from the Universidad de Concepción. Later he obtained a Ph.D. degree in Electronic Engineering from the Universidad Técnica Federico

Santa María, and was awarded the "Premio Tesis de Doctorado Academia Chilena de Ciencias 2012", for the best Ph.D. Thesis developed in 2011 for national and foreign students in any exact or natural sciences program, and is a member of the Academia Chilena de Ciencias, Chile. In the past few years, he has been a visiting professor at several international universities. He has directed and participated in several projects financed by the National Fund for Scientific and Technological Development (Fondo Nacional de Desarrollo Científico y Tecnológico, FONDECYT), the Chilean National Agency for Research and Development (Agencia Nacional de Investigación y Desarrollo, ANID), and the Paraguayan Program for the Development of Science and Technology (Proyecto Paraguayo para el Desarrollo de la Ciencia y Tecnología, PROCIENCIA), among others. He has been the responsible researcher of basal financed projects whose objective is to enhance, through substantial and long-term financing, Chile's economic development through excellence and applied research. He is the Director of the Laboratory of Energy Conversion and Power Electronics (Laboratorio de Conversión de Energías y Electrónica de Potencia, LCEEP) at Universidad de Talca, Chile. He was a full professor at the Department of Electrical Engineering at the Universidad de Talca. In April 2023 he joined the Power Electronics, Machines and Control (PEMC) research group of the University of Nottingham as a Professor. His main research areas are power electronics, renewable energies, advanced control of power converters, and microgrids, among others. He has published more than 520 academic publications in leading international conferences and journals.

M. Nageswara Rao is a distinguished Professor of Electrical and Electronics Engineering (EEE) at UCEK, JNTUK, Kakinada. He holds a B.Tech from JNTUH Hyderabad, an M.Tech from NIT Calicut, and a Ph.D. from JNTUK Kakinada. His doctoral thesis, supervised by Dr. R. Srinivas Rao, focused on "Distributed Power Generation Planning and Evaluation in Rural Distribution System under Indian Conditions-A Cost/Benefit Analysis" and was awarded in 2017. With 20 years of professional experience and 17 years of research experience, Dr. Rao has held several administrative positions including NSS Programme Officer, NSS Coordinator, Officer Incharge of Examinations, Additional Controller of Examinations, Departmental IQAC Coordinator, and Project Engineer at UCEK and JNTUK. He has guided 72 PG students and 1 Ph.D. scholar, with 9 Ph.D. scholars currently under his supervision. His publications include 16 international journals, 15 international conferences, and 15 faculty development programs. He also has a patent on a smart ring main distribution system and authored a book chapter on energy storage systems in microgrid operations. His research interests lie in power and control systems. Dr. Rao teaches various subjects, including electrical measurements, AI techniques, electrical circuits, and neural networks and fuzzy logic, and is a member of the Institution of Engineers (India). He has co-organized a workshop on AI and machine learning applications to power systems, sponsored by SERB, DST, Govt. of India.

For Product Safety Concerns and Information please contact our
EU representative GPSR@taylorandfrancis.com Taylor & Francis
Verlag GmbH, Kaufingerstraße 24, 80331 München, Germany